MEGALOPOLIS

TOURING NORTH AMERICA

SERIES EDITOR
Anthony R. de Souza, *National Geographic Society*

MANAGING EDITOR
Winfield Swanson, *National Geographic Society*

MEGALOPOLIS

Washington, D.C., to Boston

BY

JOHN R. BORCHERT

RUTGERS UNIVERSITY PRESS • NEW BRUNSWICK, NEW JERSEY

This book is published in cooperation with the 27th International Geographical Congress, which is the sole sponsor of *Touring North America*. The book has been brought to publication with the generous assistance of a grant from the National Science Foundation/Education and Human Resources, Washington, D.C.

Rutgers University Press
109 Church Street
New Brunswick, New Jersey 08901

The paper used in this book meets the minimum requirements of American National Standard for Information Sciences—Permanence of Paper for Printed Library Materials, ANSI Z39.48-1984.

Library of Congress Cataloging-in-Publication Data

Borchert, John R.
    Megalopolis: Washington, D.C. to Boston / by John R. Borchert
      p.   cm.—(Touring North America)
    Includes bibliographical references and index.
    ISBN 0-8135-1876-8 (cloth)—ISBN 0-8135-1877-6 (paper)
    1. Northeastern States—Tours. 2. Cities and towns—Northeastern States—
Guidebooks. I. Title. II. Series.
F2.3.B66 1992
917.404′43—dc20   92-9986
                                          CIP

First Edition

*Frontispiece:*The Statue of Liberty, New York Harbor. Photograph by Barbara Lerner Kopel.

Series design by John Romer

Typeset by Peter Strupp/Princeton Editorial Associates

△ Contents

PART THREE

RESOURCES

△ Foreword

Touring North America is a series of field guides by leading professional authorities under the auspices of the 1992 International Geographical Congress. These meetings of the International Geographical Union (IGU) have convened every four years for over a century. Field guides of the IGU have become established as significant scholarly contributions to the literature of field analysis. Their significance is that they relate field facts to conceptual frameworks.

Unlike the last Congress in the United States in 1952, which had only four field seminars all in the United States, the 1992 IGC entails 13 field guides ranging from the low latitudes of the Caribbean to the polar regions of Canada, and from the prehistoric relics of pre-Columbian Mexico to the contemporary megalopolitan eastern United States. This series also continues the tradition of a transcontinental traverse from the nation's capital to the California coast.

This contemporary perspective of Megalopolis, which extends from Washington, D.C., through Baltimore, Philadelphia, and New York City to Boston, examines the rich variety of environmental, industrial, commercial, transportation, and cultural phenomena. John R. Borchert, Regents Professor of the University of Minnesota, analyzes the region. He has been consulted by the U.S. government on urban problems and is an authority on the American urban system.

Anthony R. de Souza
BETHESDA, MARYLAND

△ Acknowledgments

In preparing this field guide, I have drawn on the talent and good will of many fellow geographers. Those whose names are marked with an asterisk contributed directly to the manuscript material for their respective areas. I want to express my gratitude to: Mary Lynne Bird, director, and her staff at the American Geographical Society for assistance on New York; William Casey, University of Minnesota, for assistance on Boston, New York, Washington, and many other details; Saul Cohen (*), Hunter College of the City University of New York, for assistance on New York; George Fasic, Chester County (Pennsylvania) Planning Commission, for assistance on Philadelphia; H. Briavel Holcomb (*), Rutgers University, for assistance on New York; Arthur Loeben, Montgomery Country (Pennsylvania) Planning Commission, for assistance on Philadelphia; George Lewis (*), Boston University, for assistance on Boston; Thomas Lewis (*), Manchester (Connecticut) Community College, for assistance on Connecticut and Rhode Island; David Meyer (*), Brown University, for assistance on Connecticut and Rhode Island; Judith Meyer (*), University of Connecticut, for assistance on Connecticut and Rhode Island; Arlene Rengert (*), West Chester University, for assistance on Philadelphia; John Starr, University of Maryland-Baltimore County, for assistance on Baltimore and port comparisons; Barney Warf, Kent State University, for assistance on New York; and M. Gordon Wolman (*), The Johns Hopkins University, for assistance on Baltimore. Only the limitations of time prevented me from turning to many others for their input and advice.

I also convey my thanks to Julie Tuason for an excellent job of copyediting. She was meticulous and helpful in every way. I thank Lynda Sterling, public relations manager and executive assistant to Anthony R. de Souza, the series editor; Natalie Jacobus and

Richard Walker for editorial assistance; and Tod Sukontarak for photo research. They were major players behind the scenes. I also acknowledge the work of those dedicated people who were responsible for making the maps that appear in this book: Patrick Gaul, cartographer with COMSIS in Sacramento, California, Nickolas H. Huffman, cartographer for the 27th IGC, and the following cartographic interns at National Geographic—Lynda Barker, Scott Ogelsby, Michael Shirreffs, and Alisa Solomon. Staff at the National Geographic Society's Map Library and Book Collection, Cartographic Division, Illustrations Library, Computer Applications, and Typographic Services also provided cartographic assistance to those cartographers and interns mentioned above. Many thanks, also, to all those at Rutgers University Press who had a hand in the making of this book.

Errors of fact, omission, or interpretation are entirely my responsibility, and any opinions and interpretations are not necessarily those of the 27th International Geographical Congress, which is the sponsor of this field guide and the *Touring North America* series.

PART ONE

Introduction to the Region

Megalopolis

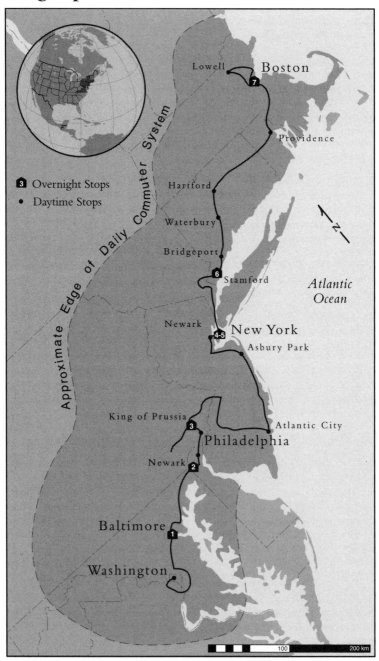

Overnight Stops

Daytime Stops

Lowell

Boston

Providence

Hartford

Waterbury

Bridgeport

Stamford

Atlantic
Ocean

Newark

New York

Asbury Park

King of Prussia

Philadelphia

Atlantic City

Newark

Baltimore

Washington

Approximate Edge of Daily Commuter System

100   200 km

△ Introduction

Megalopolis is the urban geographer's term for the urbanized northeastern seaboard of the United States. The region stretches from southern New Hampshire southwestward to northern Virginia, and from the eastern Appalachians to the Atlantic coast. It envelopes the string of the "Big Four" central cities—Boston, New York City, Philadelphia, and Washington, D.C.—and constellations of their suburbs and satellites.

Studying the region in the 1940s and 1950s, the gifted French geographer Jean Gottmann searched for a name for it. "The name," he said, "should be new as a place name but old as a symbol of the long tradition of human aspirations and endeavor underlying the situation and problems now found here." He chose to call the region *Megalopolis,* after a Greek city-state that was founded with great aspirations two and one-half millennia ago. Though the city-state abjectly failed to fulfill its founders' hopes, its name became an English dictionary term for "a very large city."

Gottmann's study became a landmark in urban geographic analysis, so the term Megalopolis found its way into the jargon of media, planning, and politics. Subsequently, the term has been applied in the United States to urban southern California, the urbanized southern Great Lakes region, and sometimes to smaller metropolitan clusters. But the northeastern seaboard, from Boston, Massachusetts, to Washington, D.C., is the prototype.

Compared with other parts of the United States, Megalopolis is a unique combination of great population numbers and density, history, wealth, physical and social diversity, and dynamism. When the nation took its most recent decennial census, in 1990, Megalopolis was home to 45 million people. One-sixth of all Americans live here, occupying about two percent of the country's land. Residential densities on the outermost fringes of the region generally exceed 100 persons per square mile and are increasing. In the old urban cores the average population densities per square mile are decreasing, but they generally exceed 10,000 and reach more than 60,000 on New York's Manhattan Island. Historically, the region was the major center of population and economic activity at the time of the founding of the United States. It was the gateway for virtually all of the country's massive stream of European immigration and the launching pad for much of its economic development. More than half of all Americans lived in the region at the time of the first census, in 1790, and in 1870 it still accounted for about one-third of the country's population.

Compared with its population, Megalopolis has a disproportionately large share of the nation's wealth, personal income, commerce, and industry. In the late 1980s its cities accounted for nearly one-fourth of the country's wholesale trade. They were home to more than a quarter of all Americans engaged in finance and law. Residents of these cities were paid more than one-fifth of Americans' total personal income. Like the region's population size, its wealth is in part a historical legacy. In 1870 it probably contained at least half of the nation's wealth.

The region's physical setting includes a large share of rugged, forested Appalachian mountains and hills, yet some of the most productive farming in the country thrives at the suburban edges of both Philadelphia and the Long Island side of New York City. The glaciers of the Ice Age left their mark profoundly on the part of the region north of the latitude of New York City. When the glaciers melted, the resulting rise in sea level drowned the edge of the continent and lower reaches of the rivers. That submergence created the region's intricate shoreline of estuaries, capes and bays, tidal marshes and lagoons, and its hundreds of miles of beaches and picturesque islands.

But the physical setting also includes a gigantic, diverse accumulation of buildings and other structures. About one-third of all the high-rise office space in the country forms the downtown and suburban skylines of Megalopolis.

The cities of Megalopolis have a disproportionately large concentration of the nation's total burden of structural obsolescence. The region contains roughly one-third of all the standing structures built in the United States before the onset of the automobile era in the 1910s and 1920s. Those buildings tend to be the most deteriorated in the country, and their design, engineering, and locations have been rendered obsolete by the technologies of the automotive age.

On the other hand, Megalopolis is a region of immense auto-era urban dispersal and decentralization. Though its population growth has slowed to a crawl since 1970, Megalopolis built new housing to accommodate 12 million residents between 1920 and 1970—greater than the growth of metropolitan southern California in the same era.

The social diversity is even more impressive. The masses of metropolitan settlement centered on New York City, Boston, Philadelphia, Baltimore, and Washington, D.C., are home to about 8 million blacks, 3.5 million Hispanics and Latinos, and 1.5 million Asians. More than five million people speak a language other than English in their homes. Along with their disproportionate shares of the nation's wealth and commerce, these places have comparably disproportionate shares of the nation's poverty, crime, and ignorance (i.e., inadequate knowledge).

And lastly, there are the region's dynamics. The massive population and accumulation of structures is a one-time inventory in a churning physical and social process.

Here, on about two percent of the nation's land, one-sixth of the U.S. labor force travels between home and work. Eight million persons change residences at least once in a five-year period. Hundreds of thousands of both English-speaking whites and minorities migrate into or out of a city or county in a single decade.

Neighborhood communities are created by the congregation of immigrants establishing a foothold within the wider society and economy. The same communities change or dissolve as succeeding generations differentiate and move away. The process of community creation and dissolution is as unending as human curiosity and striving.

Travel and shipping in Megalopolis accounts for more than one-sixth of the country's vehicle-mileage. Daily activities consume more than one-sixth of the energy used in the United States and generate an equal share of all the nation's garbage, rubbish, and waste gases and fluids.

At a different scale, people are continually converting raw material into new buildings and aging buildings into solid waste, in an unending cycle of new construction, imperfect maintenance, deterioration, replacement, and abandonment.

Of course, these are only generalities. If you go to Megalopolis and look more closely, the diversity of social and physical landscapes is breathtaking. You will be entertained, stimulated, often overwhelmed by the magnitude and complexity of the region. You perhaps will be perplexed by the persistence with which your observations challenge preconceptions and defy general or simple answers to the questions that inevitably arise. But the experience will greatly enrich your view of yourself and the world.

In fact, many of us *do* spend bits and pieces of a lifetime intensively sampling bits and pieces of Megalopolis—a monumental structure or neighborhood here, a natural or historical landmark there, museum visits sandwiched within business trips, a family pilgrimage to Washington, an excursion to New York's theater and shopping districts.

To augment that approach, a field *reconnaissance of the whole region* is worth a week or ten days in the life of every American citizen and foreign traveler. A region-wide reconnaissance can provide perspective and structure, pulling threads of coherence and continuity through the bewildering array of points of interest. The trip provides an unparalleled opportunity to observe and reflect on one's place in the historical-geographical evolution of not only the United States, but also the entire modern world.

This field guide is an invitation to just that kind of reconnaissance. The guide sketches out the geographical structure of a route along the axis of Megalopolis, from metropolitan Washington, D.C., to metropolitan Boston.

The trip provides a framework—no more, but also no less. A lifetime can be spent in fleshing it out. You can take innumerable side excursions to explore cities or valleys or bays with greater breadth, or you can make longer sojourns in particular places to

Urbanized Regions within Megalopolis

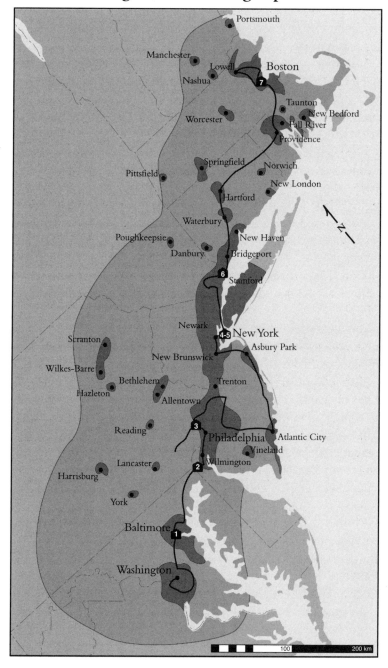

experience them in more depth. Even if you hold to the eight-day schedule, you will want to explore a bit more at some places, or pause and reflect a short while longer at others.

The route of the field trip is embedded within the larger cluster of big cities that makes up Megalopolis. The darkest shading on the map on page 7 represents the contiguous land areas that are mostly or entirely built-up in each of the U.S. Census Metropolitan Statistical Areas (MSAs) in the region.

The continuous, built-up stretch from metropolitan New Haven, Connecticut, to metropolitan Wilmington, Delaware, is home to more than 25 million population and covers one-tenth of the region's land area. Then add the big islands of contiguous urbanization around Boston, Providence, Baltimore, and Washington, D.C., with an additional 11 million people, plus the urbanized islands in twenty-four smaller Census MSAs, with an additional 6 million. That brings the total metropolitan urbanized land area to perhaps one-sixth of the region, with a population of more than 41 million.

But most of the remaining land is far from empty. Between the metropolitan built-up areas, more than 2 million Megalopolitans inhabit hundreds of small cities and villages, as well as dwellings seldom more than few hundred yards apart, along thousands of miles of country roads. Perhaps one percent of those are truly farm dwellings. The vast majority of households are supported by urban-type jobs, and in every other respect they are urban. Fundamentally, Megalopolis is a vast, interlocking urban network of job trips, shopping and service trips, and social or recreational trips linking dwellings, non-residential buildings, and recreational open spaces.

The network is dramatized vividly if you fly over the region on a dark winter morning and watch the lights come on as the region's millions of people rise and begin their day. In a short time, the buildings, roads, and some of the rail lines come alive with hundreds of millions of points of light, and tens of millions of moving lights mark the streams of traffic from home to work. They converge from a multitude of dwelling places and diverge again to several million places of work. The pattern of commuting is incredibly intense in the large metropolitan areas. But, beyond those high-intensity centers, the interconnected networks of Megalopolis branch out to the farthest edges of the region.

When you begin each day of this excursion and watch the morning rush hour traffic along your route—in every kind of vehicle and afoot—picture yourself in that swarm of 20 million people headed for their jobs. Along your route, the swarm is most dense. You can visualize it thinning gradually, fitfully outward, beyond the horizon, to the edges of Megalopolis.

Landforms in Megalopolis

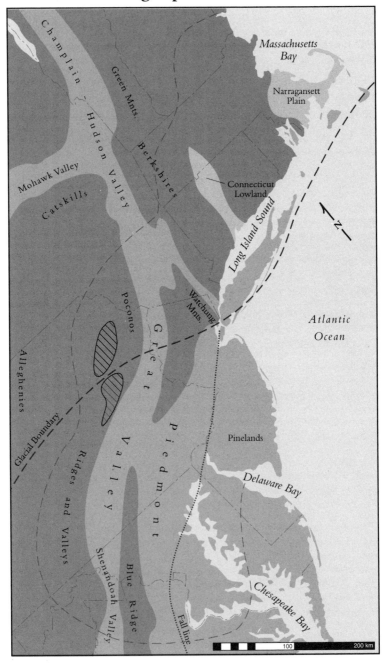

THE LAY OF THE LAND

The Megalopolis region includes a rich sampling from the natural landscapes of the Fall Line, the Coastal Plain, the Piedmont, and the glaciated terrain of the Northeast, from northern New Jersey across southern New England.

THE FALL LINE

From Washington, D.C., to the northern suburbs of Philadelphia, and again near New Brunswick, New Jersey, our route weaves along the Fall Line—a major boundary on the map of North American landforms. The higher, hard bedrock of the Piedmont rises to the northwest of the boundary. The lower, soft bedrock of the Coastal Plain drops away to the southeast.

The regional drainage system follows the general slope of the continent, from the Appalachian Mountains to the Atlantic Ocean. Hence streams generally flow from the northwest to the southeast. At the Fall Line they tumble from the Piedmont to the Coastal Plain. As a result, where rivers cross that line, early navigation from the ocean terminated, and the turbulent water was harnessed for industrial power. Thus, our route from Washington to Philadelphia, and again at New Brunswick, passes through venerable cities large and small, strung like beads at those combination port and water-power sites along the Fall Line. Likewise, our route also passes numerous Fall Line parks surrounding picturesque, rocky glens or pioneer mill sites, where you can stretch, rest, picnic, and enjoy the riverine scene.

THE COASTAL PLAIN

The Atlantic Coastal Plain slopes away gently to the southeast of the route. Soft, geologically young layers of sedimentary rock—sand, clay, and marl—dip southeastward and eventually run beneath the bed of the ocean. The relief of the land is low, and wide, flat, often marshy floodplains line the streams. Low bluffs—a few feet to a few tens of feet high—separate the bottoms from the gently sloping uplands.

Large areas of the Coastal Plain were cleared historically for agriculture. Small, diversified farms mostly disappeared long ago. But, at the hands of both local entrepreneurs and major food corporations, specialized, intensive production of vegetables, fruit, flowers, nursery stock, milk, and poultry has evolved and survived. Commercial crop farming is generally found on the outer margins of the commuter zone, where land is least expensive. Meanwhile, small farms have experienced a rebirth in the hands of middle- and upper-income urban commuters who use the land or rent it for grazing small herds of beef cattle or horses. Elsewhere, trees are slowly reclaiming the land.

Our detour from Philadelphia by way of Atlantic City and Asbury Park in New Jersey crosses part of the Coastal Plain. Between Philadelphia and Atlantic City the route nicks the edge of the distinctive Pinelands (also known as the Pine Barrens) of central New Jersey. The soft sedimentary rocks there are virtually pure sand. Because of the infertile, excessively drained soil, agricultural settlers generally avoided the area. Pine woods and savanna continued to dominate the landscape. Meanwhile, the growing coastal resorts to the east pumped a large and increasing amount of groundwater from the sedimentary sands deep beneath them—the same sand layers that surface under the Pinelands to the west. Coastal communities came to depend on soak-in on the

Pinelands to recharge their wells, and protection of the recharge areas from latter-day land development is an environmental issue.

At the coastline, the layers of Coastal Plain sediments dip beneath the Atlantic Ocean. The short landscapes testify to the unstable relationship between land and sea. They remind anyone who looks carefully that land uses near the shore are subject to natural forces which operate at the human time scale but with energy that dwarfs any human ability to control them.

Low terraces stand tens or scores of feet above present sea level. They commemorate the relatively recent geologic time, before the last ice ages, when the sea stood high enough to flood a substantial part of today's outer coastal lands. Some coastal bluffs are comprised of deep layers of the shells of clams that were direct ancestors of those which inhabit offshore waters today.

Broad estuaries form the lower reaches of rivers and streams. Most prominent are those of the Potomac and Susquehanna rivers (Chesapeake Bay) and the Delaware River (Delaware Bay). When the ice-age glaciers temporarily locked up a small part of the world's ocean water, and sea level fell several hundred feet, those streams deepened and extended their valleys out to the receding shoreline. But when the glaciers melted to their present extent, the rising sea level drowned those same lower reaches to form today's deep harbors, brackish-water marshes, and tidal flats. At the same time, waves on the rising sea pounded the coast and eroded the headlands. Shore currents swept up the freshly eroded sediment and strung it out to form long, straight sandbars that partially blocked the bays, formed quiet lagoons and salt marshes, and built broad, dune-lined beaches facing the open ocean.

Most of the land is soft—the bluffs, tidal flats, beaches and dunes. Waves and tides and rainstorms are endlessly modifying the lay of the land. And the power of a "northeaster" in winter or a hurricane in autumn speeds the processes of change, often catastrophically.

Plant life, marine life, bird life, and eventually human settlements evolved in this dynamic coastal setting. It is the setting for the Atlantic coastal flyway, the fisheries, and the developed shores of Megalopolis.

THE PIEDMONT AND BEYOND

The Piedmont province of the Appalachian Highlands rises northwest of the route. Long, gentle slopes connect rolling uplands to broad valleys that wind several hundred feet below the ridge tops. Sweeping views abound from openings in the rich hardwood forest. Weathering through geologic time has formed moderately deep soils on the underlying ancient, hard, crystalline rocks.

Agriculture developed early in the Piedmont and has endured. The richest and most famous of its farming districts extends westward from Philadelphia toward the gap in the mountains between Reading and York in Pennsylvania. Lancaster County, Pennsylvania, is the heart of the district, and Lancaster city is the county seat. One spur of our route reaches into that county just east of the city of Lancaster. Like the Coastal Plain, the Piedmont has also become the location of many clusters of urban-based, part-time farms.

Southwest of York and northwest of Reading, 500- to 1,000-foot-high walls of folded and fractured crystalline rock mark the western edge of the Piedmont. Southwest of York, and on to the west of Washington, the wall is the famous Blue Ridge. Northeast of Reading, it is a less clearly defined collection that includes the Schooley Mountain of northern New Jersey, the landmark Watchung Mountain ridges that rise west of Newark, New Jersey, and the Ramapo Mountains along the northwestern suburban fringe, on the New York-New Jersey boundary. The same highland mass curls east and south from the Ramapos to include the spectacular Palisades, which tower above the west bank of the Hudson River in the northern New Jersey suburbs opposite Manhattan.

Between Reading and York, the Piedmont opens westward into the Ridge-and-Valley province of the Appalachians. A belt of old sedimentary rocks—severely folded and uplifted in the geologic revolution that formed the Appalachians 200 million years ago—stretches from Alabama northeastward into Canada, and part of it underlies some of Megalopolis. Layers of rock thousands of feet thick were tilted so they stand vertically, like the pages in books on a shelf. Eons of erosion cut down the layers as they were uplifted and folded, so now the vertical hard rock-beds stand as long, northeast-southwest trending mountain ridges, and the softer rocks lie as broad, rolling valleys.

In the eastern part of the province, most of the rock is soft, and valleys dominate over ridges. The area has long been called the "Great Valley" of the Appalachians. Inland metropolitan cities of Megalopolis are strung along the Great Valley from Poughkeepsie, New York, to York and Harrisburg in Pennsylvania. The Great Valley continues, west of Washington, as the famed Shenandoah Valley of Virginia; it runs northward from Megalopolis, as the Hudson-Lake Champlain Lowland, between the Adirondack Mountains (New York) and Green Mountains (Vermont), into Canada. Like the Piedmont, the rolling lands of the Great Valley were opened to agriculture early in American history.

Still farther west in the Appalachian province, ridges are dominant and valleys are narrow and discontinuous. The country is also higher and rougher. But the intense urban development that eventually coalesced into Megalopolis was attracted into part of that higher, rougher country by the rich coal deposits of the Pennsylvania Anthracite Region. Those coal deposits were the basis for a major nineteenth-century mining industry focused on the metropolitan area of Scranton-Wilkes Barre-Hazelton and for industrial development at trade centers in the Great Valley—Bethlehem, Allentown, Reading, Lancaster, York, Harrisburg.

Beyond the Ridge and Valley, the outer edges of the Megalopolis circulation overlap the rugged, deeply dissected Appalachian Plateau, in the Allegheny, Catskill, and Pocono mountains.

Like the Fall Line that separates Appalachian uplands from Coastal Plain lowlands, another major physical boundary trends generally east-west across Megalopolis: the southern limit of continental glaciation during the ice age. The glacial boundary, which lies roughly at the latitude of Manhattan, was muddled and diffused in this area by the extremely uneven terrain over which the ice sheet was moving. Nevertheless, the glaciers left a profound mark on the natural drainage pattern. In the unglaciated area, the stream pattern is regular and completely developed, from the broad, low main valleys up to the smallest tributaries and gullies that gnaw at the narrow ridge-top divides. But in the glaciated area, slopes of the valley walls are softened, valley bottoms are partly filled with glacial deposits, and uplands are partly smoothed. The drainage pattern is also irregular: lakes, ponds, and swamps result from disturbance of the well-developed system of mainstreams and tributaries that had evolved in the eons preceding the ice age.

THE NORTHEAST

The Piedmont-Coastal Plain boundary goes to sea beneath Long Island Sound. The Coastal Plain is represented only by Long Island, the Narragansett Plain of southeastern Massachusetts, and the tiny but significant Boston Basin. All show the effects of glaciation. Hummocky, bouldery ridges (moraines) stretch east-west across Long Island, marking the southernmost advance of the glaciers in that part of the world. Moraine deposits also appear on Block Island, Martha's Vineyard, and Nantucket, off the coast to the east of Long Island. The moraine deposits help to make the north shore of Long Island distinctly higher and more irregular than the south shore. Meanwhile, the low, sandy glacial outwash deposits, spread by meltwater south of the ice sheet, give the south shore its distinctive flat terrain, long, straight beaches, and shallow lagoons. Wave action on the glacial deposits has also produced the long beaches of the Cape Cod National Seashore in Massachusetts.

The rest of Megalopolis north of Long Island Sound and east of the Hudson River is in the glaciated Appalachian uplands. Faulted and folded blocks of crystalline rock, with structures trending generally north-south, give the whole region its distinctive topographic grain. The Green Mountains of Vermont and White Mountains of New Hampshire give way southward to the Berkshires in western Massachusetts, the hills around Wachusett Mountain (elevation 2,006 feet) in eastern Massachusetts, and the still lower ridges running southward across Connecticut and terminating in coastal headlands. Thin glacial deposits are scattered over the rocky uplands.

Deposits are thicker in the valleys and allow for locally important beach developments where the valleys meet the sea.

The first seven or eight generations of European-Americans farmed the least inhospitable tracts of this rocky land because they had little choice. Where they did not clear the forest for crops and pasture, they cut it for firewood. But most of the crop and pasture land has been in the process of gradual abandonment since the 1830s, and firewood slowly shifted from a necessity to an urban luxury after about 1870. Hence, nature has gradually restored the rural woodlands, and an extremely rich and varied woodland it is. Two factors in the environment encourage that variety: the great range of slope and drainage conditions in the rugged, glaciated terrain; and the extremely steep climatic gradient, from the off-shore waters of the warm Gulf Stream to the upper-middle latitude interior highlands. One of the world's profound winter climatic boundaries separates the foggy, mild offshore islands from the snowy interior of the New England portion of Megalopolis.

The Connecticut River valley occupies the most impressive of the north-south trenches in northern Megalopolis. From Spring-field, Massachusetts, south to New Haven, Connecticut, the trench widens to form the distinct lowland area centered on Hartford, where significant crop agriculture survived longest in the face of general New England land abandonment.

THE RIVERS

Water is about the only commodity in which Megalopolis is self-sufficient. On the average the region produces at least as much as it needs—in fact, probably much more than it needs. But there are some remarkably big management requirements.

Storms and air masses from the tropical Atlantic deliver forty to fifty inches of precipitation in an average year. Direct evaporation, along with transpiration from the vegetation cover, take up about two-thirds of that. So about one-third remains to run from the uplands into the lakes and rivers and eventually back to the sea.

From myriad creeks and springs, the runoff gathers eventually into a few major rivers. Longest and most voluminous is the Susquehanna. It delivers the equivalent of more than one-third of the runoff of all Megalopolis into the head of Chesapeake Bay. The Connecticut is about half as large; the Potomac, Delaware, and Merrimac perhaps one-tenth. The Schuylkill, Passaic, Housatonnic, and Blackstone are smaller still. But, like the others noted here, those short rivers have played exceptionally important roles in both the economic development of the nation and the urban development of Megalopolis.

Runoff and stream flow from the land of Megalopolis amount to about fifty billion gallons daily—five times the average daily requirement for combined domestic, commercial, and industrial uses. Hence, most cities have been able to meet their requirements for pure water by building reservoirs on small streams nearby, and buying and controlling the use of land in the watershed. Washington and Philadelphia have pumped directly from the large rivers at their doorstep during most of their history. In the twentieth century New York City has built long aqueducts from reservoirs in the Catskills, and Boston has reached west to the slopes of Wachusset Mountain. But even in those cases, the reservoirs impound relatively small upper tributaries. You will see examples of the impressive reservoir system of Megalopolis on our traverse through the region.

Thus, on the face of it, no water supply problem is apparent. On the other hand, the supply of natural runoff is variable. In a major drought period it is one-fifth of the average; hence, the supply is barely adequate, and shortages develop during periods of heaviest use. In addition to the one-fifth of the natural runoff used by cities, commerce, and manufacturing, another three-fourths or more is required to cool thermal-electric generating plants. To allow for low flow, their supply of cooling water must be recycled.

The supply also becomes heated and contaminated from very heavy use. More than 90 percent of all the used water is eventually returned to the rivers. However, the return from municipal systems is partly treated sewage, and the returned cooling water has been heated. Furthermore, about one-tenth of the runoff

Rivers

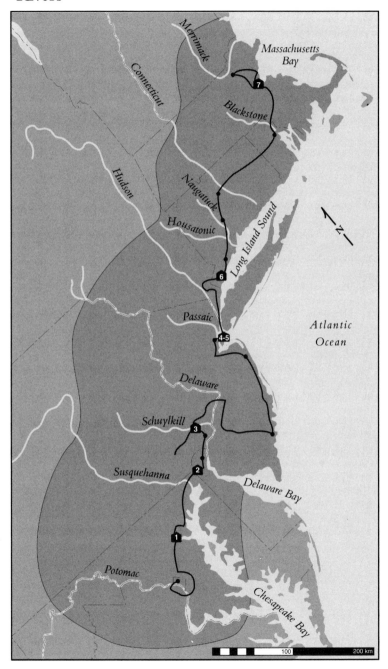

comes from urban streets and storm sewers, which contribute to the contamination.

So the region's water supply is no longer ample if it is used exploitatively. Continued or expanded use requires continued, expanded investment in storage, treatment, and watershed controls.

SETTLEMENT AND ECONOMIC DEVELOPMENT

At the limits of each city or village along the route through Megalopolis is a small sign that declares the date of municipal incorporation. Established Megalopolitans take the dates for granted. Southerners from Virginia to Georgia, or from the Bluegrass or Nashville Basin, find the dates fairly commonplace. But visitors or immigrants from the rest of the country are fascinated by the age of these places.

Of all the central cities of metropolitan areas along the route, only Baltimore and Washington have founding dates more recent than the 1600s. Baltimore's establishment came rather late in the settlement of the Chesapeake Bay area. And the settlements at Georgetown and Alexandria—now part of metropolitan Washington—preceded the settlement at Washington by half a century.

Many plaques along the route also testify that the initial settlers were a heterogeneous lot—different religious beliefs, different attitudes toward race and human relations, Dutch preceding the British in New York and the Delaware River valley, Swedes preceding the Dutch and British in Wilmington and the Delaware River valley. Although the initial thrusts were imperial and military, most of the settlers were motivated by the opportunity to pursue a living and to escape from some kind of tyranny.

Founding Dates of Today's Major Cities

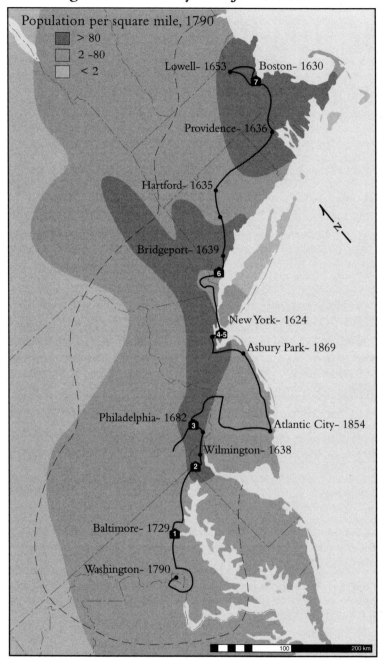

Population per square mile, 1790
- > 80
- 2 – 80
- < 2

Lowell– 1653   Boston– 1630

7

Providence– 1636

Hartford– 1635

Bridgeport– 1639

6

New York– 1624

4-5

Asbury Park– 1869

Philadelphia– 1682

3

Atlantic City– 1854

Wilmington– 1638

2

Baltimore– 1729

1

Washington– 1790

N

100   200 km

EMERGENCE OF THE REGION

From the earliest European settlement until about 1920, the ports of Megalopolis were the gateways for settlers and their commerce. The ports were a "hinge" between developed western Europe and developing Anglo-America. On the western side of the Atlantic, they were the indispensable transfer points between water and land in the corridor that grew to link urban-industrial northwestern Europe with the unmatched natural resources of the Great Lakes waterway, Lake Superior ores, Appalachian coal, and level, fertile plains of the Middle West.

In the resulting hemispheric stream of traffic, the port cities were, fundamentally and simply, critical locations where people competed for the opportunities to do business. A substantial part of the heterogeneous stream of travelers and immigrants put down roots at these places. They lived, worked, traded, and organized communities and institutions to serve their purposes. The local governments, and eventually the nation itself, were some of those organizations. The settlers also built the structures they needed to shelter their activities and connect the different locations they occupied. Once in place, those structures were destined to last a century or more. Replacement would be negligible for several decades, and then always done gradually. Some of the buildings in the cities were monuments to important people and organizations, built to stand "forever."

The settlement pattern quickly acquired an inertia. The structures and the organizations and the locations became locked into one another. To be sure, for nearly three centuries—until about 1920—growth at these places mainly reflected their location in the world's overwhelmingly predominant and intensifying circulation corridor from northwestern Europe to the American Manufacturing Belt and Corn Belt. Meanwhile, however, the inertia of the cities of Megalopolis was becoming an important characteristic in its own right.

By the time of the first United States census in 1790, two million Americans resided in what is now Megalopolis. They comprised more than half of the country's population, and carried on an even larger share of its economy. Isolated colonial settlements—initially congregated on the basis of religious beliefs, language, and national background—had coalesced to form an almost continuous belt of high population density from Boston to Baltimore. Population had grown to more than 100,000 in each of the clusters of counties that now comprise metropolitan Boston, New York, and Philadelphia. The density of more than 80 per square mile was the highest in Anglo-America and among the highest in the western hemisphere. It equaled the densities in the outer commuter reaches of Megalopolis today, and it was already about one-third non-farm.

In a way, the population pattern on the first census map resembles today's map. From New York City south to northern Virginia, today's western boundary of Megalopolis was, in 1790, the quintessential western frontier of European-American settlement in the nation, though, of course, the Spanish had permanent settlements in the Southwest by the 1590s and the French were settled in the Lake Superior region by the 1620s.

NERVE CENTER AND MIRROR OF NATIONAL GROWTH

Much less quaint and much more conspicuous than the historic plaques along the route through Megalopolis are the massive accumulations of structures built in the nineteenth and early twentieth centuries. A large number of them are respectably maintained. A small share are handsome reminders of the best architecture and most powerful people and institutions. A large number also have reached advanced stages of neglect and deterioration—tenements, downtown department stores, railway stations and sprawling yards, docks, multi-storied warehouses and factories, even the public streets, schools, and parks in those districts.

That mass of structures is the physical record of the Railroad Era, from the 1830s to the 1920s. The structures represent the emergence and growth of the great industrial metropolis in American historical geography. Most of the structures from earlier times were replaced in the process of building cities during this era.

Anchor for the Initial Westward Movement

Commerce and migration through the port cities of Megalopolis grew as the United States expanded westward from the Atlantic seaboard. From the time of the first census in 1790 to the 1820s, the frontier west of these cities shifted from the crest of the Appalachians all the way to Illinois and the population of the country tripled from 4 million to 12 million.

Abundant free or cheap land for farming was the big attraction. But the settlers went to the frontier to carry on commercial agriculture, not for the endless drudgery of subsistence farming. They knew about the commercial economy of the North Atlantic rim, and they expected to be a part of it. Likewise, people in the established cities on the East Coast expected their commercial interests to benefit from the expansion to the west.

Hence, the pressure for transportation and communication improvements was intense. Roads were one obvious response. Widened, graded routes for wagons and stagecoaches reached westward from tidewater at Albany (New York), Philadelphia, and Baltimore. But the roads represented no basic change in technology; vehicle speed and capacity were still the same, so transportation costs rose in direct proportion to distance.

Canals were a more innovative, though temporary, response. They reflected engineering advances in the form of locks, aqueducts, storage, and diversion. They offered the cheaper alternative of water transportation to reach inland locations, as well as salable industrial water power sites. Major canals appeared on the map of the urbanized northeast in the 1820s and 1830s.

Most famous and important was the Erie Canal, begun in 1817 and completed in 1825, which linked the Hudson River with Lake

Erie. By opening a water route from the Atlantic to the Middle West, it gave New York City an unbeatable advantage in the competition for international trade. Competing attempts to reach the deep interior by canal were the Pennsylvania system, which reached westward from Philadelphia, and the Chesapeake and Ohio Canal, from Washington up the Potomac River into western Maryland. Among numerous shorter canals, one served industry along the Blackstone River in Massachusetts and Rhode Island; another ran beside the Delaware to within reach of the anthracite coal fields in Pennsylvania; and a later network reached up the Susquehanna into southern New York.

Headquarters for Developing the Economic Core

The first railroads appeared around 1830 and opened a new era. The technology for the "iron horse" evolved quickly, and soon trains combined speed and capacity to render both roads and canals obsolete for medium-distance movement of cargo, passengers, mail, and express. Long-haul and heavy, bulk cargo continued to move mainly by water. Hence, the early rail network was complementary to the main inland waterways—the Great Lakes and Western Rivers (Ohio-Mississippi-lower Missouri system)—and to the long-standing coastwise ocean routes that linked the main port cities (Baltimore, Philadelphia, New York, Boston).

An outgrowth of the Railroad Era was the development of the country's first national transportation system by 1870. A dense network of interconnecting rail lines and major natural waterways focused on the great ports of the mid-Atlantic and Northeast, the Great Lakes, and the Western Rivers. The network covered the quadrangle from Boston and Baltimore-Washington on the east to Minneapolis-St. Paul and Kansas City on the west. It was a combined rail-water system, slow and creaking by later standards, but it embraced and developed the American Manufacturing Belt and Corn Belt—regions which would be the economic core of the western hemisphere for the next century.

Railroad Era Transportation Routes

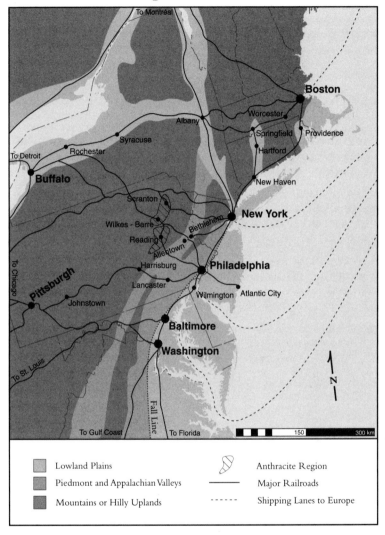

Lowland Plains

Piedmont and Appalachian Valleys

Mountains or Hilly Uplands

Anthracite Region

Major Railroads

Shipping Lanes to Europe

The development of that vast network resulted in the emergence of a new generation of major locations for potential decentralization of commercial and political power. Chicago, St. Louis, Cincinnati, and Pittsburgh were leading examples. There were new opportunities to shift from the older, established great ports of the northeastern seaboard. Indeed, decentralization was one result of westward expansion.

But the spread of the economic core westward brought commensurate growth back in the emerging Megalopolis. Cities there were historically the places of initial accumulation of capital. The commercial families in those places had most of the connections with sources of capital in western Europe. They had strong family, business, and professional ties with most of the educated entrepreneurs who went west.

As a result, much development capital, influence, and control remained in the eastern cities. They continued to grow with the nation as a whole, and people there continued to provide much of the direction for building the rest of the country. In fact, throughout the century of the Railroad Era, and for another half-century up to 1970, the urbanized northeastern seaboard persistently accounted for about 20 percent of the population growth of the United States.

Climax of the Railroad Era

The growth of a modern steel industry in the United States after the Civil War opened a new chapter in the rail era. In a decade, from 1870 to 1880, the country shifted from an importer to an exporter of steel. The entire existing main-line railroad network in the economic core region was relaid with steel rail. Locomotives became bigger and more powerful, cars heavier, trains longer and faster. The different gauges of the northeastern network became standardized. Other innovations followed—automatic couplers, air brakes, blocking systems, refrigerated cars. The speed and capacity of the system took a major leap upward, and the heyday of the general-purpose river, lake, and coastal packet ships drew to a close. Domestic water transportation could only specialize in the heaviest or highest-bulk cargoes—notably coal, oil, and grain.

The rail network of the American Manufacturing Belt and Corn Belt could now be expanded to form an integrated national—even continental—system. Elaboration of the piecemeal collection of lines in the South could bring new opportunity for industrialization and economic integration of that region. Transcontinental railways could now open up the Great Plains, the intermountain oases, and the Pacific valleys, ports, and gateways to Asia. New major cities could begin to take root.

Direction for the task came primarily from the reservoir of capital and talent in the economic core region in general, and the northeastern seaboard in particular. The influence of Megalopolitan banks, brokerage firms, law firms, transportation and communications organizations, manufacturers, wholesalers, publishers, and entertainers reached out to the far ends of the new national, standardized rail network. Meanwhile, at the southern end of the emerging Megalopolis, Washington lay waiting for an occasion to bring the national government more actively into the management of this new, huge, complex national system.

The map of main railroad corridors in 1920 reflected the historic position of the urbanizing northeastern seaboard. Atlantic shipping had evolved from the technology of small sailing ships to clippers to big liners and now concentrated on the major harbors with the widest rail connections, at Philadelphia, Baltimore, Boston, and especially New York City.

Rail networks radiated to the interior from those centers, reflecting the dreams and efforts of empire builders at each place. The New York Central main line reached up the Hudson River valley, across the Mohawk-Ontario lowland, then westward beyond Buffalo. The Pennsylvania main line ran from Philadelphia to Pittsburgh and westward. The Baltimore and Ohio main line—the shortest route from tidewater to the Midwestern plains—snaked westward through the mountains to Pittsburgh and the middle Ohio River valley.

The anthracite roads reached east and west from their coal-field base to offer a cross-mountain link from both Philadelphia and

New York City to Buffalo and the Middle West. With its lines reaching southward along the Piedmont and Coastal Plain, Washington was the major gateway to the South. Lines up the Merrimac and Hudson rivers reached to Montréal from Boston and New York. A master link—the Northeast Corridor of today's Amtrak trains—joined Boston and Washington. That corridor superceded the post roads and coastal shipping lanes that historically joined the northeastern ports. It continued to mark the location of the "hinge" in the larger European-American hemispheric circulation system.

With the immense growth of commerce and wealth, geographical signs of affluence and leisure followed in short order. Incidental to their commercial missions, the rail lines opened up valuable access points to long-romanticized scenic and recreational resources. Lines from Boston passed through the White and Green mountains and the Berkshires. Besides their economically important link to the Middle West, the Buffalo lines went to Niagara Falls, the Catskill and Pocono mountains (respectively in New York and Pennsylvania), and the glaciated Finger Lakes region of upstate New York.

On New Jersey's 150 miles of ocean beaches, the potential recreational traffic alone was great enough to justify some new construction. Not long after the Civil War, a line was built to the beaches at the point nearest Philadelphia, and Atlantic City, New Jersey, was born. By the 1880s a network of lines from New York City and Philadelphia reached the beaches at many points between Asbury Park and Cape May on the Jersey shore.

Nearer the cities, the main lines ran through neighboring open countryside. Along those routes—especially outside New York, Boston, and Philadelphia—new suburbs were created beginning in the mid-nineteenth century. Local branch lines, as well as the main lines, were also pressed into service around all three cities.

But the expansion of the contiguous, high-density areas of the established big cities along spreading networks of streetcar lines made the new rail-connected suburbs look trivial by comparison. In emerging successive rings, builders put up vast new tracts of wall-to-wall row houses and tenements, town houses, and closely spaced single-family homes on forty- or fifty-foot lots.

In each new generation, households in the growing middle class tended to occupy the newer housing toward the edges of the cities. This left the older tenements and tiny row houses to "filter down" to lower-income people, many of whom were immigrants arriving in massive streams from Europe and who settled in crowded ghettos that were partly congregated and partly segregated. Some of the areas evolved into well-kept, distinctive neighborhoods as particular ethnic groups prospered and remained. Others deteriorated into slums, as particular ethnic groups or individuals prospered and then migrated.

Thus, a swelling stream of humanity—generation after generation—moved through a burgeoning stock of aging buildings and dispersed into a burgeoning stock of new ones. Basic processes subtly grew to unprecedented proportions—aging and obsolescence of structures, geographical concentrations of poverty, segregation, neighborhood turnover and succession, disintegration of communities, kaleidoscopic organization and reorganization. At any time the scene appeared simply as physical, social, and political turmoil. Our traverse of Megalopolis samples landscapes inherited from all of these epochs.

The Automobile Era

One vivid memory from your trip through Megalopolis will be the traffic congestion and the noise. Frustration comes easily in a stream of thousands of vehicles converging and diverging among multiple lanes. Stop-and-go delays are frequent. Missing a sign and getting lost is a real risk, although it will almost surely get you into interesting neighborhoods you would have missed otherwise. Noise envelops you—from traffic, from planes overhead, from blaring car radios.

An equally vivid impression will be the hundreds of square miles of single-family homes and apartment buildings in the distinctive styles of the 1920s and post-World War II building booms, as well as the hundreds of suburban commercial strips, retail malls, motels, office and industrial parks. All of those landscapes reflect two important facts: the population of Megalopolis doubled, from 20 million to 40 million, between 1920 and 1970; and that half-century was the age of the internal combustion engine, cheap oil, and the maturation of the broadcast industry.

In a way, there was nothing new in these developments. The urbanized northeastern seaboard continued its century-old tradition of accounting for 20 percent of the national population growth. But there were some very important new elements. Highway and air transport offered unprecedented mobility. The coast-to-coast broadcast networks—together with the maturation of the motion picture industry—brought a new wave of information about far-away places. The new wave of mobility and information was reinforced during World War II.

Much of the population increase between 1920 and 1970 was accounted for by the post-World War II baby boom. That was by far the largest generation of Americans who would grow up entirely in the age of automotive and broadcast culture and technology. How would they think about the question of where to live when they entered the job and housing markets?

Much of the population increase was accounted for by migration from the Deep South—especially by blacks. With the stream of migration from eastern and southern Europe cut to a trickle after World War I, sustained growth in the Automobile Era came to depend more on natural increase and limited out-migration of people already living in the region, and on black immigration from the South.

The cities of the northeastern seaboard had long had significant black populations. Slaves comprised 30 to 50 percent of the population in the Washington and Baltimore areas at the time of the first census in 1790. Thousands of free blacks lived in the Philadelphia and New York City areas before the Civil War.

Overall, the percentage of blacks throughout the region was rather small in 1920, but by 1970 Megalopolis was home to one-sixth of the black population of the country. The region's population was more than ten percent black—concentrated in the central cities. Racial and cultural divisions, as well as sheer numbers, increased and exposed the turmoil that had already accompanied the process of succession, neighborhood turnover, and physical deterioration in the rail era.

In this same half-century, from 1920 to 1970, the railroad network lost perhaps half of its importance. Intercity rail passenger traffic became negligible. Short-and medium-distance trips shifted to the highways, long trips to the air. Rail freight rose in terms of ton-mileage, but the railroads specialized in bulk, high-volume and long-haul cargo in the face of a massive increase in trucking. Rail companies scaled back operations, using far less trackage, equipment, and labor. Branch lines were virtually eliminated.

The interstate highway network replaced the main-line rail network as the major, general-purpose land transportation system. On the face of it, the patterns of the two networks in Megalopolis look much alike. They serve the same major cities, and the density of the network is greatest in the Boston-Washington corridor.

But a closer look shows some important differences. The rail network was dominated by a few sets of radials focused on the four great port cities: Boston, New York, Philadelphia, and Baltimore. In contrast, the freeway network resembles a grid that provides more direct connections between each large- or medium-sized city and any of the other cities. There are more options and more flexibility. New major routes increase the accessibility of areas that were formerly outside main transportation corridors. There are more locational options for business, employment, and recreation.

The census metropolitan areas appear as huge halos around the much more compact rail-era central cities. In the automobile age, while the population of Megalopolis doubled, the urbanized land area multiplied by four or five times. The explosion of numbers of vehicles and mileage of good roads put large areas of rural land within the urban reach. Land all around the metropolitan fringe became relatively cheap, and central-city densities fell. Within range of any workplace, most people had more options of where to live.

Very large international airports appeared on the map beyond the edges of the rail-era central cities. From each of the great old

Freeways and Metropolitan Areas in the Automobile Era

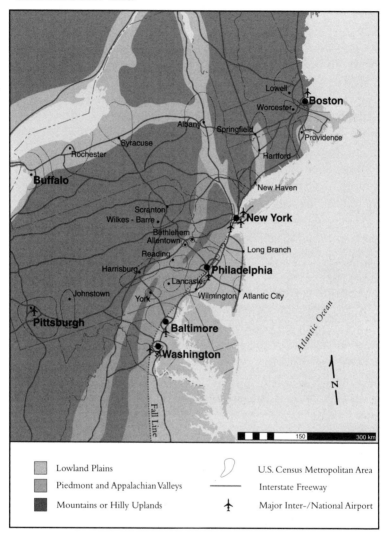

Lowland Plains

Piedmont and Appalachian Valleys

Mountains or Hilly Uplands

U.S. Census Metropolitan Area

Interstate Freeway

Major Inter-/National Airport

ports, direct, global intercity linkages multiplied and accelerated. The locational advantages for any business near the intercity terminals had shifted from the rail-era downtown district to the airport district, since most airport districts were better served by freeways.

Meanwhile, the freeway system has mostly bypassed the railroad-era city cores. The central cities appear on the maps as big lumps embedded in the metropolitan area—mostly older structures and many of them obsolescent. Those lumps could be penetrated by new routes or rebuilt only at great cost.

By the 1970s, it was becoming evident that the settlement pattern of Megalopolis was in the midst of a massive and irreversible realignment. Hundreds of billions of dollars had been spent on a new pattern, adapted to the auto age. And many more billions would be spent before the new expansion would be fully mature. Meanwhile, hundreds of square miles of structures had been rendered obsolescent. Abandoning, razing, or rehabilitating those areas—and somehow modernizing their access routes—could require additional, unscheduled billions of dollars.

There was also a dawning sense that in some important ways changes in transportation and communications since 1920 might have diluted the role of Megalopolis. Compared with the early twentieth century and before, accumulations of wealth and human skills had become more widely distributed, both in the United States and in the world. Opportunities had become more dispersed among a larger number of metropolitan regions. People in each of those regions had gained more direct access to all others, without having to go through Megalopolis.

Yet, despite these changes in its strategic position in the Automobile Era, the region continued to account for its traditional one-fifth of national population growth. How important had sheer inertia of organizations and structures become in accounting for that continued growth? Many analysts tended to dismiss the new opportunities for wider choice and flexibility that had come with new transportation and communication technologies and the emergence of important new metropolitan centers. They emphasized instead the continuity of giant organizations based in Megalopolis, and the continuing need for face-to-face contacts among multitudes of employees and small entrepreneurs to carry on their business. Meanwhile, the question persists.

In the 1960s, when the nation was preparing to land a person on the moon yet the suburban rail service into Manhattan was in shambles, humorist Art Buchwald wrote of a mock interview with a fictitious Dr. Haven Hartford, who was in charge of a project to land a commuter from Connecticut on Manhattan Island. After a lengthy discussion of the problems, he asked Dr. Hartford, "Why do we really have to land a man on Manhattan, anyway?" In a famous quotation of the time, Haven Hartford replied, "Because it's there!" His answer underscored the inertia of the place. Perhaps it applied to much of Megalopolis.

The Information Era

Since 1970, the Megalopolitan share of national population growth has dropped from 20 percent—its stable level for nearly two centuries—to 4 percent. Black immigration has slowed sharply, white net out-migration has accelerated, and birth rates have followed the national decline. Those trends have been only partly offset by the new wave of immigration from Latin America and the Caribbean, Asia, Africa, and eastern Europe. Emigrants have been attracted to other regions of the U.S. by growing economic and educational opportunities, and by perceived improvements in physical or social environment. Immigrants have come to Megalopolis for the same reasons, of course; however, the balance has shifted significantly.

Washington, D.C., is an exception, however, to the overall decline in the Megalopolitan share of national population growth. Although the growth rate of Washington has slowed down since the 1970s, it continues to exceed the national rate and it still reflects the federal government's centralization and spectacular expansion since the 1930s.

An explosive growth and spread of information has accompanied what I call the jet-satellite-computer age that began in the

1970s. The baby-boom generation entered its age of maximum mobility, in Megalopolis as elsewhere, and a new wave of information rolled over the Third World. The range of known options of places to which to migrate attained unprecedented scale, and the number of people ready to migrate exploded both at home and abroad. Little wonder that more people left Megalopolis for the rest of the country, and more arrived from the Third World.

The information revolution brought a comparable explosion of locational options for many kinds of industries, businesses, and institutions. Enterprises and organizations continued to move into Megalopolis or to spawn there, and also to move out or to spawn elsewhere, but the balance shifted elsewhere.

In the face of these changes, the local governments and other locally-based community organizations faced a new problem.

Similar to medieval gates on the Rhine River in Germany, the cities of Megalopolis had long occupied virtually unavoidable gateways in the European-American corridor. Within fairly wide limits it was possible to exact comparatively high fees for doing business there—in all facets and corners of the total political, social, and commercial system. Historically, the traffic had few alternatives to paying up.

Few would deny the comparatively high cost of doing business in Megalopolis. For example, municipal general revenue and expenditures have long been high. In the late 1980s, New York City and Washington, D.C., spent four to six times as much per capita as did Chicago and Los Angeles. The other large cities of Megalopolis spent one-and-a-half to two times as much.

During the new information era the relative location of the region's cities has changed, and their communities face unprecedented levels of competition. Communities in other locations compete for parts of the global traffic into and out of the eastern United States. Labor forces elsewhere compete to apply their skills to virtually every task performed in the export economy of Megalopolis. Unavoidable pressures for efficiency and change confront entrenched monopolies or oligopolies at these historic commercial locations.

Effects of this loss of centrality and monopoly have been most pronounced in the central cities. Masses of old buildings were abandoned at a faster pace as the shift to suburbia sped up and the inflow of poor immigrants slowed down. For the most part, the residual population tends to be the least mobile—hence, also the poorest. One important result is a decline in the residential property tax base, a loss that has been offset partly by an increase in the construction of offices and hotels. Local governments, with federal aid, have subsidized the construction with the hope of further shoring up the tax base, but both the market and the availability of subsidy money are limited. Many major cities, and states, are thus operating in bankruptcy or on the edge of it.

The result in the central cities by the beginning of the 1990s was a very mixed landscape. The downtowns displayed towering new skylines and considerable cosmetic improvement, but they were mostly surrounded by large areas of deteriorated buildings and public facilities, poor inhabitants, and vacant land. And the future was uncertain in many outlying neighborhoods. There seemed to be a growing impatience to confront festering questions objectively. Yet clear, coherent courses of action were evasive, as they have always tended to be in this unpredictable, complex system.

Adjustment to a changed position for Megalopolis in the global system proceeds, but in various directions. As you traverse the region, you will see symptoms: in the urban cores, of increased emigration, of new open space, of thinning densities, of rehabilitation and new structures, of maintenance and clean-up; in both the urban cores and on the sprawling fringes, of fill-in, of maturation, and of a shift from exploitation toward rational management. But you will also see symptoms of new polyglot streams of immigrants from teeming remote regions of the world, with stressful acculturation, as well as symptoms of inadequate information and understanding, greed, neglect, abandonment. You will see symptoms of technologic and social innovation alongside dogged adherence to irrational, outmoded routines and structures. You can conjure numerous, diverse future scenarios from combinations of these trends.

Megalopolis continues to be "a very large city." As you traverse the region, you will wonder whether it—and the nation—are a system or a contraption. You will wonder just what the differences are across the gray zone between symphony and cacophony. But as you develop a greater understanding of the region—of its greatness as well as its blemishes—you will no doubt come to agree with Jean Gottmann who, over three decades ago, wrote the following: Megalopolis is ". . . a symbol of the long tradition of human aspirations and endeavor underlying the problems now found here."

Megalopolis remains a unique region of the North American continent, and one every serious traveler seems destined to explore. The itinerary that follows, in Part II, is an attempt to help you understand all aspects of the region from a geographical point of view.

P A R T   T W O

The Itinerary

THE ROUTE AND THE STRATEGY

The traverse of Megalopolis begins at the Washington Monument, which is the heart of the Mall in the nation's capital. The itinerary includes a full day or more in metropolitan Washington, Baltimore, Philadelphia, and Boston. It includes more than two full days in metropolitan New York. Another day is filled by traversing central New Jersey to the Atlantic City boardwalk, then to more representative developments on the shores of Megalopolis. And yet another day is given to the traverse from New York City to Boston, with time to sample certain highlights of Hartford, Connecticut, Providence, Rhode Island, and the "quiet corner" of Megalopolis in northeastern Connecticut. An evening probe westward from Philadelphia reaches into the Amish country and rich agricultural "island" of Lancaster County, Pennsylvania.

The field notes (i.e., chapters) in Part II are grouped by metropolitan areas as defined by the U.S. Census Bureau, not by day-segments of travel. Except for one county in northeastern Connecticut, our route is everywhere within a census Metropolitan Statistical Area (MSA). Five of those areas, with a combined 1990 population of 30 million, lie "wall-to-wall" from Washington, D.C., through New York City. Between metropolitan New York and metropolitan Boston, and on the detour to the New Jersey coast, the route crosses the smaller metropolitan areas of Atlantic City, New Haven-Waterbury (Connecticut), Hartford, and Providence.

U.S. Census Metropolitan Statistical Areas along the Route

Boston

Providence

Hartford

Waterbury

New Haven

● Urban Centers

◇ U.S. Census Metropolitan Statistical Area Boundaries

◇ County Boundaries

Atlantic Ocean

New York

Philadelphia

Atlantic City

Baltimore

Washington

100   200 km

Thus, separate sets of notes cover metropolitan Washington, Baltimore, Philadelphia-Wilmington (Delaware Valley), Atlantic City and the Jersey Shore, New York City, central Connecticut and Providence, and Boston. In each area, the route is designed to sample both the variety of landscapes and the variety of processes that have produced them: rolling uplands and plains; areas that are monumental and ordinary, new and old, maintained and neglected, rich and poor; areas for living, working, and recreation; planned developments and spontaneous ones; areas of congregation and segregation; white, black, and mixed-race neighborhoods; English-speaking and non-English-speaking areas; teeming streets, lush farms, and empty enclaves.

Now, take the field guide and maps in hand, and spend eight days playing a tiny part in the greatest crowd scene on the North American stage!

◸ Day One

METROPOLITAN WASHINGTON

Prologue

The Census Metropolitan Statistical Area (MSA) comprises the District of Columbia and ten counties in Maryland and Virginia. Metro population was 1.6 million in 1950, 3 million in 1970, and 3.4 million in 1984. Blacks represented 23 percent in 1950, 24 percent in 1970, and 26 percent in 1984. Metropolitan per-capita personal income is the highest among urban areas in the country.

The Fall Line divides the District and the metropolitan area roughly into northwestern and southeastern halves. The deep valley of the Potomac River separates the Maryland and Virginia sectors of the Piedmont; and the scenic Rock Creek gorge bisects the Piedmont portion of the District. The wide, tidal, lower Potomac separates the Maryland and Virginia parts of the Coastal Plain.

The major rail routes northeast to New York City and south to Richmond, Virginia, follow the inner edge of the Coastal Plain and reinforce the Coastal Plain-Piedmont contrast. The routes northwest to Pittsburgh, Pennsylvania, and southwest to Atlanta, Georgia, climb the Piedmont. In the auto-air age, there has been little freeway penetration of the Railroad Era city. The congested, circumferential Capital Beltway is the inner boundary of most of the radial, intercity freeway system. The heavily-subsidized, showcase Metro subway system is designed to provide access to the core. National Airport—one of the two busiest in Megalopolis—is inside the Beltway but is served by both Metro and freeway. The historic central business district of Washington lies between the Capitol and the White House, off Pennsylvania Avenue. Following serious deterioration in the post-World War II years, it has been the scene of considerable recent redevelopment. Meanwhile, the central office district has spread far northwest and west to the Potomac River. Large suburban office and retail centers have grown around the ends of some Metro subway lines and near some major interchanges on the freeways.

Residential development before World War II took place mainly inside the District. Extensive subsequent growth has been accommodated outside the District. Higher residential land values are found in the western and northwestern sector, on the Piedmont. Values tend to be lower in the eastern sector, in the rail-industry corridor, and beyond, on the Coastal Plain. Black population is strongly concentrated in the northern and eastern District and suburbs.

Public open spaces are extensive, associated with both the affluence and the historic national importance of the area. They are predominantly in the Piedmont sector. Although the basic street layout of the central city follows the famous plan of French engineer Pierre-Charles L'Enfant (b. 1754; d. 1825), and buildings in the monumental core follow a roughly coordinated scheme, most development has been spontaneous and incremental. Nevertheless, a few small, comprehensively planned enclaves reflect ideals and efforts of particular developers or cultures at particular times.

Guideposts along the Route

THE MONUMENTAL CITY

To begin the trip head eastward from the Washington Monument on the Mall. This two-mile-long open space and its surrounding buildings are the core of monumental Washington—architectural symbol of the nation, principal mecca for millions of tourists each year, and breathing room for thousands of Washingtonians. It's also the location of 95,000 federal jobs—one-fourth of the metropolitan-area total. The exceptionally large share of employment in the central city helps to justify the showpiece Metro rail transit system. Within ten miles are the headquarters for agencies that employ more than 3 million people worldwide and spend the equivalent of one-fourth of the national income.

The monumental buildings along our route on the right, in succession, are the Department of Agriculture, Freer Gallery of Art and the buildings of the Smithsonian Institution (the Castle, the Arts and Industries Building, the Hirshhorn Museum and Sculpture Garden, and the National Air and Space Museum).

The U.S. Botanic Garden lies south of the Capitol Reflecting Pool. A second layer of buildings is visible in the background, south of Independence Avenue and east from the Department of Agriculture—the Department of Energy, the Federal Aviation Administration, the Department of Education, and the Department of Health and Human Services.

At the eastern end of the Mall, swing northward around the Capitol. The House of Representatives office buildings stand to the south, the Library of Congress and Supreme Court to the east, the Senate office buildings and the Department of Labor building on the north. Perhaps the most important part of this complex is the Library's Geography and Map Division—one of the world's great map libraries, where any visitor is welcome.

Returning westward from the Capitol grounds, look northwest up historic Pennsylvania Avenue, through the redeveloping Railroad-Era downtown, toward the White House and the extensive blocks of offices to the northwest. But continue west along the north edge of the Mall. You will pass, in order, the National Galleries of Art, the National Museum of Natural History, and the National Museum of American History. In the background, north of Constitution Avenue, you glimpse the Labor Department, the Federal Court House, the National Archives, the Justice Department (the fastest-growing major agency in recent years), the Internal Revenue building, the Interstate Commerce Commission, and the Department of Commerce.

Skirt the north side of the Washington Monument and continue westward on Constitution Avenue. The White House appears be-

yond the Ellipse, between the Treasury and the Old Executive
Office Building, which used to house almost the entire Washington
federal bureaucracy when it was built after the Civil War. The
Organization of American States and the National Academy of
Sciences retain their shade trees and lawns, while the massive
State Department building looms behind them.

The Vietnam Veterans Memorial and the Lincoln Memorial
mark the end of the Mall as you turn north on 23rd Street. As you
leave the monumental city core, look toward the Potomac River
from 23rd Street and glimpse the John F. Kennedy Center for the
Performing Arts and the giant Watergate complex—one of the
largest mixed-use developments in the country and the site of the
embarrassment that led to the resignation of President Richard
Nixon in 1974.

Traverse a few blocks of the Foggy Bottom district and sense
the extremely high (and increasing) residential density and continuing redevelopment in the close-in, fashionable quarters of the
city. Pass some of the buildings of George Washington University, then follow Pennsylvania Avenue and M Street into historic
Georgetown.

Pause to walk along the Chesapeake & Ohio Canal locks below
Wisconsin Avenue south of M Street. The hand-powered locks
accommodated barges that were scarcely larger than a modern
pickup truck and drawn by horses on the tow-path. Begun in the
1820s, the canal was outmoded by the railroads as a link to the
interior long before it was eventually completed in 1850. The
Georgetown locks were part of the descent of the waterway from
the level above the falls of the Potomac River to the level of
tidewater. The drop also provided a limited amount of water
power. Contrast the colonial vintage row-houses fronting the waterway with monumental Washington. Equally contrasting, note
the early nineteenth-century industrial and commercial buildings
located on the canal for both accessibility and power and now
rehabilitated for office and residential use.

Rehabilitated early nineteenth-century row-houses and 1970s replacement luxury apartment buildings in the Foggy Bottom district of Washington, D.C., have led to increased density and increased land values. The old row-houses typify much of Georgetown.

GEORGETOWN TO ALEXANDRIA

Cross the Potomac River into Virginia on the Key Bridge. Note the buildings of Georgetown University (founded by the Jesuits in 1789) rising above the Potomac behind you to the right, and head south among the office, hotel, and residential towers of Rosslyn. Like numerous large-scale commercial developments outside the District boundaries, this one reflects the relief of extreme pressure for space in the post-World War II expansion of the federal government, the opportunity to avoid District of Columbia building-height restrictions, and the location of major suburban highway junctions.

Like those in other similar centers, Rosslyn's towers are occupied by many independent federal agencies and many consulting and lobbying firms. These are a small fraction of the thousands of federal offices and related firms that also inhabit miles of twelve-

Metro Washington

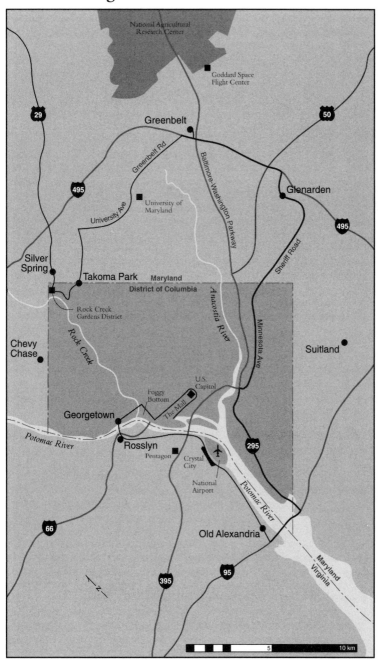

National Agricultural Research Center

Goddard Space Flight Center

Greenbelt

Greenbelt Rd

Baltimore-Washington Parkway

University Ave

University of Maryland

Glenarden

Sheriff Road

Silver Spring

Takoma Park   Maryland

District of Columbia

Rock Creek Gardens District

Rock Creek

Anacostia River

Minnesota Ave

Chevy Chase

Suitland

U.S. Capitol

Foggy Bottom

The Mall

Georgetown

Potomac River

Rosslyn

Pentagon

Crystal City

National Airport

295

Potomac River

66

Old Alexandria

Maryland Virginia

N

395

95

5   10 km

Residential and office towers of Crystal City, near Washington National Airport, along U.S. Highway 1.

The "Condo Canyons" of the newer highway corridors of Alexandria, Virginia. Industrial parking lots in foreground are part of the rail corridor leading to Richmond and the South.

The rehabilitated colonial facades of Old Town Alexandria.

story block fronts and rehabilitated townhouses in central Washington, as well as the contemporary buildings set in office parks along the Capital Beltway.

Continue southward from Rosslyn past the mansion of Gen. Robert E. Lee (b. 1807; d. 1870) and the lawns of Arlington National Cemetery on the edge of the Appalachian Piedmont. Then proceed past the world's largest office building, the Pentagon—location of an important share of the Defense Department's nearly 90,000 civilian jobs in metropolitan Washington; administrative headquarters for nearly 1 million civilian jobs worldwide; headquarters for administration of 25 percent of total federal expenditures, for 16 percent of all governmental expenditures at all levels in the United States, for 5 percent of total national and local government expenditures in the world (all as of the early 1990s). Take a second look at the Pentagon as you pass by—a rather modest building considering its importance in the system of global accounts!

On to the south pass Crystal City, west of the George Washington Memorial Parkway. A massive office-retail-residential megastructure, it has many functions similar to Rosslyn's plus several million square feet of high-rise residential space.

To the east of the parkway sprawl the well-maintained terminal facilities and runways of Washington National Airport—one of the nation's busiest, oldest, and most cramped airports, because of lawmakers' insistence on keeping domestic flights within easy reach of the Capitol. International flights and a few long-distance domestic flights use the newer Dulles Airport, located 25 miles to the west.

The transportation theme continues as you pass the Potomac Yard of the rail lines that lead to Richmond, Virginia, and south—a reminder that Washington, D.C., is also the head of the rail corridors that extended northeastern industrial and banking influence through the southern states to Florida and the Gulf Coast between 1870 and 1920.

Then, suddenly, you enter the Old Town preservation district of Alexandria—colonial port at the head of the Potomac estuary. Outside the District, on the Coastal Plain rather than on the Piedmont, beyond the railroad yards, and farther from central Washington, Alexandria lagged behind Georgetown by perhaps a decade in entering the present age of gentrification. Don't pass through Alexandria without roaming a few of the side streets off Washington Avenue. Read the historic plaques. Visit the Torpedo Factory near the waterfront—first of many "festival" shopping malls you'll encounter in restored corners of Megalopolis. Visit Gadsby's Tavern, where George Washington often met with friends. And pick up a copy of the weekly newspaper, the *Gazette-Packet,* established in 1734. Its masthead alone tells you something of both the size and the function of this city in colonial times. Look also at the news items and the real-estate advertisements. Together they reflect the place of Alexandria astride the blighted rail-industry corridor and public-housing areas, the burgeoning condominium canyons and row-house projects that line the arterial highways radiating across the Coastal Plain, the high-value traditional single-family homes that spread downriver near the Poto-

Washington, D.C., to Central Baltimore

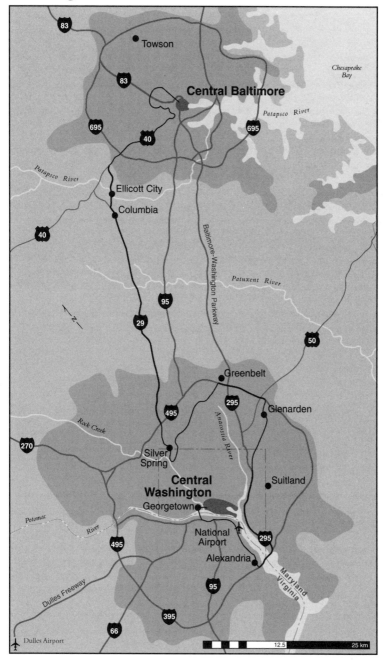

A new row-house development in a predominantly black, east-suburban area of Prince Georges County, Maryland.

mac toward Mt. Vernon, and the pricey historic restorations of Old Town.

INLAND COASTAL PLAIN

Over the Beltway's Woodrow Wilson Bridge, on the opposite side of the Potomac from Alexandria, exit at Interstate 295 (Anacostia Freeway), and follow the southern shore of the Anacostia River's tidal estuary, where it cuts across the southeastern part of the District of Columbia. Military installations, whose origins go back to the early nineteenth century, line the lower Anacostia. To the north, toward the Capitol and the Robert F. Kennedy Memorial Stadium, on the horizon, lie some of the major poverty and ghetto districts of Megalopolis. Four miles to the east, up the low bluffs just outside the District boundary, lies the sprawling United States

Census Bureau complex at Suitland, Maryland, where thousands labor thanklessly to try to comprehend the almost incomprehensibly complex nation of which Washington, D.C., is the capital.

Leave the freeway at the 11th Street Bridge and follow Minnesota Avenue and Sheriff Road northeastward. You'll pass through yet a different Washington—the historic blue-collar, one- and two-family, small-home part of the city oriented to the streetcar lines of the Railroad Era. It retains much of its physical character, although today it is a mostly black, low-income neighborhood.

Sheriff Road leaves the District of Columbia near its eastern corner and enters Prince Georges County, Maryland. In the past twenty years, as the growing population of the predominantly black eastern precincts of Washington has spread further eastward, Prince Georges has become the most populous predominantly-black suburban county in the country. While its total population of about 700,000 has grown only slightly, the black share has grown from 14 percent to more than 60 percent since the 1960s.

Glenarden and its vicinity provide a good place to see some of these changes written on the land. Between Sheriff Road, the Beltway, and the last two stations on the Metro's Orange Line, Glenarden's first streets and lots were platted from a plantation in 1910. It was an isolated, impoverished, rural black community. By the late 1930s the combination of New Deal reforms, the growth of federal employment, and the automobile had brought the place into the Washington daily orbit. It was one of the first three predominantly black towns in Maryland to be granted a charter by the legislature—all of them in Prince Georges County. Suburbanization came rapidly after World War II. A city of about 5,000 in 1990, and annexing more developing land, Glenarden is impressive for its community organization, its quality of public infrastructure, housing, and general maintenance, and the diversity of ages and incomes in its population.

Within a two-mile radius are the Capital Centre Arena—home of the Washington professional basketball and hockey teams—modest Beltway office parks and industrial parks of varying quality, a regional shopping mall and numerous commercial frontages ranging from run-down pre-World War II collections to new strip malls, tidy new town-house developments among green lawns on curvilinear streets, and public-housing blocks whose social disorganization, crime, and drug traffic are a source of concern in Glenarden. These are Washington's typical eastern suburban landscapes, mixed and scattered over the broad, wooded divides, retreating farmlands, and gentle slopes of the Coastal Plain.

The scene stands in strong contrast with the monumental, globally oriented core of Washington, only ten miles west. But the two are linked by regional transportation and utility networks and the network of payrolls that branches ever outward from the massive core of federal employment.

From Glenarden follow the Martin Luther King, Jr., Highway (Route 704) northeast to Glen Dale Road (Route 193), and northwest on Route 193 to Greenbelt. The route passes one corner of the Department of Agriculture's National Agricultural Research Center. The National Aeronautics and Space Administration's Goddard Space Flight Center is nearby. (Note: The former is a world center to improve the world's growth of food and fiber. The latter is a world center to improve the observation and mapping of how much is grown where.) Beyond those two tracts lie the Patuxent Wildlife Research Center and Fort Meade military reservation—altogether, more than forty square miles stretching halfway to Baltimore.

A short distance to the west of those research centers, leave the Beltway once more, to visit the city of Greenbelt. Built by the New Deal Resettlement Administration in 1936 as a model community for low-income families, Greenbelt is known worldwide as a landmark in the history of urban planning.

Of a 12,000-acre tract ceded from the National Agricultural Research Center, the original town used 217 acres. The remainder comprised a surrounding green belt from which the place took its name. The original 885 dwelling units were open to families in the $800–2,200 annual income range. Eligible families were screened from thousands of applicants on the basis of their interest in a cooperative community and potential leadership. The residents were not low-income by national standards; acceptable income range was actually in the middle half of the family income spread

In planned communities designed to reduce the need for automobiles, residents may have trouble finding space to store the household vehicles. A scene in Greenbelt, Maryland.

for the country. The housing was not low cost—36 percent higher per unit than the average new dwelling built in the country that year. The rents were subsidized to more than compensate for the relatively high cost of development.

There have been some big changes over the years. The federal government violated the green belt in 1941 with the construction of a defense housing project of 1,000 dwelling units. In 1953 the government sold the project to a cooperative of the residents but only after selling three-quarters of the remaining green-belt land to private developers. When the resident coop leadership then found itself in financial difficulty because of the withdrawal of federal subsidies, more of the remaining open land was sold for development. By 1990 the population had grown from the original 2,800 to almost 18,000, and the remaining open areas were filling with major retail, office, and hotel structures. The Greenbelt museum helps to recall how things originally were meant to be. Like planned and controlled residential communities generally, Greenbelt has stood the test of time. Well-maintained, with a strong sense of community, it remains a popular place to live for an above-average-income segment of the residential market. It also illustrates the unplanned environmental changes that can drastically change the plans for even planned communities.

Greenbelt Road and University Boulevard lead westward, across the westernmost of the rail-industry corridors in the eastern Coastal Plain portion of the metropolitan area, past the campus of the University of Maryland, College Park (one of the country's major universities and intercollegiate athletic powers, founded in 1807), and back into the Piedmont.

THE PIEDMONT

Leave University Boulevard and follow Flower Avenue and Carroll Avenue through Takoma Park—a community of predominantly owner-occupied single-family homes established by the Seventh Day Adventist religious denomination in the late nineteenth century. The denominational Columbia Union College campus (founded in 1904) is a reminder of the early development, but today's population is rich in ethnic and racial diversity, though mainly in upper income and education brackets.

Pick your way westward into the District of Columbia, past the grounds of famous Walter Reed Army Medical Center, to 16th Street. Notice that you are following the eastern edge of the valley of Rock Creek. Bordered by one of the nation's great urban parks, the creek plunges southward through a deep, winding gorge to the Potomac River in Georgetown. The breaks along the gorge were an important attraction for the high-value northwestward residential expansion of the city in the late nineteenth and early twentieth centuries.

Detour briefly through the side streets west of 16th Street into the Rock Creek Gardens neighborhood, in the extreme northern corner of the District. Single-family homes, typically valued at one-half to one million dollars (at 1991 prices), were built in the

A high-value residential area on the Piedmont in northwest Washington, on an unusual winter day when two inches of wet snow paralyzes traffic. The spire of the National Cathedral appears in the background.

1920s, but expensive new ones—designed to look vintage—were added on large lots in the 1980s. This neighborhood provides a taste of the landscapes that dominate the Piedmont upland from here southwestward in the District, westward through suburban Chevy Chase and Bethesda, in Maryland, on to the northwestern quadrant of the Beltway, and—with lower densities and newer homes—far beyond the Beltway into the lush, rolling countryside. In these Piedmont neighborhoods, median household income is two to three times the figure for the eastern, Coastal Plain suburbs.

Now pick up Route 384 (Colesville Road) northeastward for a few blocks into downtown Silver Spring—another late-nineteenth-century suburb with a major retail, office, and high-rise residential building boom—begun after World War II and enhanced since it became a terminus of the Metro system's Red Line in the 1970s. Like Rosslyn and the others, the buildings here house many newer and non-cabinet federal agencies, business-service companies, and consulting firms.

If it's near a mealtime, get directions to Crisfield's seafood restaurant for a gastronomic reminder that you really are in Maryland, plus a look at the homespun, one- or two-story concrete-block look of commercial Silver Spring before World War II. Then return to Colesville Road (now U.S. Route 29), and head northward.

Exurban subdivisions, rural estates, and part-time, hobby, and specialty farms line Route 29 for twenty miles. Midway to Baltimore you cross the Patuxent River. A short way downstream, the river crosses the Fall Line at Rocky Gorge Dam. About seven miles farther along Route 29, the buildings of Columbia, Maryland, peek through the trees. This is Howard County—in the Baltimore metropolitan area by U.S. Census Bureau definition, but within the overlapping commuter zones of both Baltimore and Washington. The county population was 40,000 in 1965, when its commissioners approved developer James Rouse's plan to create a new city on 14,000 acres of land. A quarter-century later the county had grown to nearly 200,000 people, and the new city of Columbia exceeded 110,000.

The city was laid out to be an isolated, self-contained collection of nine "Villages," with a wide range of dwelling types and densities, around a "Town Center," with clustered commercial buildings and spacious landscaping. Several small reservoirs graced the rolling Piedmont landscape. The plan also provided for extensive industrial- and office-park developments on the fringe. Village and street names epitomize the contrived "history-and-quaintness" style of much of post-World War II suburbia. A cultural island in 1972, Columbia carried the county Democratic presidential primary for George McGovern (senator from South Dakota at the time), while the state went for George Wallace (governor of Alabama at the time). Nowadays the grass-roots Columbia Association—which has taken over community organization and promotion functions from the Rouse Corporation—observes that Columbia is "still more black, more Jewish, and more liberal than its neighbors," but older and less transient. Meanwhile, the city and county demographics are converging.

The planned edges of Columbia are also gradually merging with the rapidly multiplying, spontaneous developments in surrounding Howard County. But its community association and its design probably will make Columbia physically and socially distinctive for decades to come. Like Greenbelt and Takoma Park, it will stand as a monument to one particular effort to inject comprehensive planning and community structure into the vast, tumbling process of Megalopolitan growth and change.

Heading northward from Columbia, you not only cross the fuzzy divide between the Washington and Baltimore "commutersheds," but also enter a different history and a different economy—that of metropolitan Baltimore.

Baltimore Metropolitan Area

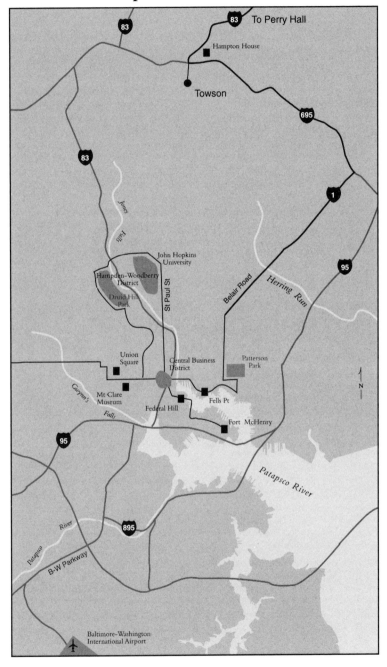

△ *Days Two and Three*

METROPOLITAN BALTIMORE

Prologue

The Census Metropolitan Statistical Area (MSA) of Baltimore comprises six Maryland counties and the independent central city of Baltimore. One of the counties, Queen Annes, is on the eastern shore of Chesapeake Bay, tied to the rest of the metropolitan commuter network by the Annapolis Bridge. Population increase since World War II has been much slower than that of Washington, D.C., and like the rest of Megalopolis. MSA population was 1.5 million in 1950 and 2.1 million in 1970. In the subsequent slow-growth years, it had increased to 2.2 million by 1984. Blacks represented 19 percent of the MSA population in 1950, 24 percent in 1970, and 27 percent in 1984. Black share of total population growth jumped from 34 percent between 1950 and 1970 to 71 percent between 1970 and 1984.

The Fall Line trends northeast-southwest through the MSA. In the city of Baltimore and the counties to the northeast, land near the Chesapeake Bay is on the Coastal Plain. Anne Arundel County (which includes Annapolis) and Queen Annes County are on the Coastal Plain. Virtually all of the rest of the MSA is on the Piedmont. The areas along the Bay are fragmented by long tidal estuaries and peninsulas. The lower Patapsco River estuary forms the spacious harbor of Baltimore. In the Baltimore area numerous short rivers rush from the Piedmont to the coast. Besides the Patapsco are Gwynns Falls, Jones Falls, Herring Run, and

Gunpowder Falls. (Note: The latter is a major source of Baltimore's water supply.)

The New York City-Washington, D.C., rail corridor follows the upper edge of the Coastal Plain. Lines to the interior twist into the uplands along the valleys of the Patapsco, Gwynns Falls, and Jones Falls. Today's circumferential Beltway encircles the city and splits into three different harbor crossings in the southeast quadrant. A half dozen intercity radials fan out from the Beltway. Three partly penetrate the inner city core but do not interconnect. Historic Charles and St. Paul streets form a corridor northward from downtown onto the Piedmont.

The central business district developed at the site of original settlement at the head of Patapsco Bay—the Inner Harbor. Since the 1960s it has been the scene of a succession of clearance, redevelopment, and rehabilitation projects which now show impressive results. The harbor dominates the Railroad Era industrial scene. Vast dock and yard areas reflect Baltimore's locational advantage nearest the Appalachian bituminous coal fields and the Middle West in the Railroad Era. Smaller historic industrial corridors follow the rail route to the interior. Extensive auto-era industrial developments cluster near the Beltway.

The largest of the suburban retail and office centers is at Towson, where the Beltway intersects the major northern arterial streets. The newest lies northwest of the Beltway and is named for the neighboring, once-rural settlement of Owings Mills.

Residential development up to World War II was within the city, with higher-value residential areas generally on the Piedmont and particularly in the northern sector. Post-World War II development has been almost entirely outside the city, mainly within Baltimore County. The burgeoning black population is expanding through the central city, mainly in the north and west.

Wall-to-wall row-houses account for a remarkable amount of the residential stock—in most socioeconomic strata and through most of the history—except for the upper end of the market in post-World War II suburbia. Vivid ethnic districts persist within the central city.

Public open spaces are extensive and mainly on the Piedmont. Except for Druid Hill Park (home to the Baltimore Zoo), they tend to follow the valleys.

Guideposts along the Route

WESTERN APPROACHES

A few miles north of Columbia, Maryland, leave Route 29 and follow the sign marking the old road down the steep walls of the Patapsco River gorge into eighteenth-century Ellicott City. Grain mills at the Fall Line served neighboring colonial wheat-growing areas on the Piedmont—the farmlands around those quaint village centers you passed in today's sea of exurbia just off Route 29. Large concrete silos and mill buildings are remnants of later growth of the industry at the same site.

On the original National Road (now Route 40) to Cumberland, Maryland, and the Ohio River, Ellicott City became the western terminal of the Baltimore and Ohio (B & O) Railroad in the early 1830s. It was virtually a half-day trip from the city then, half an hour commuting time today—a difference reflected in the visible improvement of many old and decrepit buildings. An example is the rehabilitation and matching new construction in the old company settlement of Oella, just to the left of the bridge after you cross the Patapsco.

Climb from the valley back to the upland, and follow Route 40 through extensive middle-value, auto-era neighborhoods toward central Baltimore. The bridge crossing provides a glimpse of the wide valley of Gwynns Falls, with its typical park and its rail route to the interior.

THE INNER CITY

Turn south on Monroe Street (the junction with U.S. Highway 1), into the Harlem Park neighborhood. Behind the tight, street-facing facades of nineteenth-century row-houses were inside blocks, which faced the alleys and housed slaves and poor people. Public funding in the 1960s encouraged razing of the interior blocks and

Three-story houses on ten-foot-wide lots reflect the limited mobility of the pedestrian cities such as Baltimore. The poorly-maintained playground in foreground reflects present problems of infrastructure and community. The scene is near the Mount Clare Railroad Museum.

rehabilitation of the exterior buildings. The attempt to provide a nucleus for privately-financed neighborhood improvement has partially succeeded.

A short distance away is attractive Union Square. Famous journalist and satirist H. L. Mencken (b. 1880; d. 1956) lived there at the turn of the twentieth century, even when he had to commute to New York City to edit *American Mercury*. Adjoining row-houses tend to be well-maintained, and a few have been gentrified. The nearby Holland Street Market is one of several surviving small, open farm markets, dating from the turn of the twentieth century, when the area was near the edge of the city and residents beyond the edge were farmers and market gardeners rather than daily commuters to urban jobs.

East of Union Square, Pratt Street passes the B & O Railroad Museum, in the buildings and remaining tracks of the post-Civil War locomotive repair shops and roundhouse. Allow plenty of time to read the many plaques and to absorb the total panorama of railroad technology, beginning with the *Tom Thumb,* which pulled the first steam-powered trains in the United States. The exhibits will solidify your sense of the relation of port development to development of the western hinterland. Remember that Baltimore entrepreneurs bet on the new railroad technology rather than on the canal. Reaching Cumberland in western Maryland ahead of the C & O (Chesapeake and Ohio) Canal, their B & O Railroad quickly became the dominant link to the West from Chesapeake Bay tidewater.

Follow Martin Luther King, Jr., Boulevard northward past the sprawling Maryland state office building complex, then slip into the Bolton Hill neighborhood on the edge of the Piedmont. The area's fine nineteenth-century homes had deteriorated as emergency defense housing during World War II. But private investment in restoration and maintenance has returned to the area, accompanying public and nonprofit organizational investment in streets, schools, the University of Baltimore, the Maryland Institute of Art, Lyric Opera House, and Meyerhoff Symphony Hall.

Swing northwest past Druid Hill Park, inaugurated in 1860 and subsequently improved by architect and landscape gardener Howard Daniels. An outstanding urban open space in its own right, the square-mile park includes an 1860s public water system reservoir impounded by one of the country's oldest large earth-fill dams—a national civil engineering landmark. The Baltimore Zoo is also located here.

Continue east from the park through the Hampden-Woodberry area, where factories and associated worker neighborhoods first developed using the water power of Jones Falls in 1806. The valley's boom followed the Civil War; it was the major producer of cotton duck for sails of the Baltimore Clippers. Pass the Johns Hopkins University's Homewood campus—a major and esteemed private institution, founded in 1876. (Note: The Baltimore Museum of Art is adjacent to the JHU Homewood campus; admission is free on Thursday.) Then head south on St. Paul Street in the

Charles Street, near the Johns Hopkins University's Homewood campus and the Baltimore Museum of Art. Row-houses on the Piedmont reflected the higher land values on higher ground.

north-south arterial corridor. Pass the monumental Pennsylvania Railroad passenger station (it stands atop a tunnel through the edge of the Piedmont). On Amtrak's Northeast Corridor, the station is still very active and helps to anchor a small but lively festival-type commercial district extending westward toward Bolton Hill.

THE REDEVELOPING DOWNTOWN

Farther south you reach the northern edge of the central business district (CBD) at Mt. Vernon Place. The nation's oldest monument to George Washington—a tall column and statue erected in 1821 —is bordered by fine turn-of-the-century office buildings, stately townhouses, the Peabody Conservatory (part of the Johns Hopkins University), and the renowned Walters Art Gallery.

At the southern edge of the CBD, go to the platform at the top of the World Trade Center office tower for a fine panoramic view of

A colonial church and new office towers, at the western edge of the Inner Harbor renewal area in central Baltimore.

New towers have replaced deteriorated, abandoned warehouses, and frame the small core of 1910s office towers in central Baltimore. First the port of Baltimore, the Inner Harbor is now used entirely for recreation and tourism.

the city. If you have kept to a rather leisurely schedule, you'll reach the tower with plenty of strong daylight and back-light for the view over the eastern quadrant of the metropolitan area and the spacious harbor. Plaques and brochures help you to identify many districts. Major dock and industrial districts sprawl toward the Sparrows Point steel plant, on the eastern horizon. Nearby are historic Camden Station—now the terminal for the second commuter rail line to Washington, the new major-league sports stadium (Camden Yards-Orioles Stadium), the mammoth, landmark nineteenth-century B & O warehouse, and the National Aquarium. Early twentieth-century towers project among the gleaming new skyscrapers in the downtown to the north. The Piedmont upland silhouette is broken by the skyline of the Towson suburban business complex on the northern horizon.

From your downtown hotel you can use the evening to explore and relax with the crowds (if it's the warm season) in the immaculate festival shopping, dining, museum, and recreational boating facilities around the Inner Harbor. When you do, remember that in the age of the Baltimore Clippers this was one of the busiest and most prosperous commercial junctures in Megalopolis (chaotic and littered, like all of them, to be sure); and during the 1950s it was one of the most deteriorated, squalid, rat-infested corners of the region.

THE VENERABLE PORT

Begin the next day (Day Two) at the historic park on Federal Hill, commanding a fine panorama of the CBD and the Inner Harbor, back-lit in the morning sun. Read the interesting plaques and learn why it is called Federal Hill. Later, ask a random selection of Baltimoreans and see how many know the answer.

Restoration continues in the Fells Point district of Baltimore.

Continue two miles southeast to Fort McHenry. More than a square mile of rail yards, tracks, and docks for coal, grain, and general cargo sprawl on either side of Fort Avenue as you near the fort. From the shore at the fort, you can see another five miles of docklands for coal, military cargo, automobiles, and containers. Altogether, forty miles of industrial and dock facilities line the shores of the Bay. Notwithstanding its impressive history and current capacity, the port is under severe competition from Norfolk, Virginia, and ports to the south, as well as New York City, as shipping technology and global linkages change.

The fort is a national shrine because its flag, during a British naval bombardment in the War of 1812, inspired Francis Scott Key (b. 1779; d. 1843) to pen what would become the words to the U.S. national anthem. In addition to the harbor views, don't miss a stroll in the fortification, the plaques, the museum exhibits, and the immense supplemental knowledge to be gained from the park rangers. Reflect on the battery of sixty cannon and the few regiments who were defending this piece of the continent. Today, U.S. peacetime military expenditures per capita, adjusted for inflation, are about twenty-five times as great as they were at the peak of the War of 1812. One possible interpretation: it is less expensive to harass a world power than to be one.

Return from Fort McHenry to the CBD waterfront, and proceed eastward through transitional areas to surprising Fells Point. Here an alliance of neighborhood property owners, initially resisting the invasion of a freeway, stimulated a major renewal effort. The subject is an early nineteenth-century collection of waterfront warehouses, blocks of stores and houses, and a public market for farmers who brought their produce by boat on the Bay.

PERSISTENT ETHNIC NEIGHBORHOODS

Proceed eastward through the adjacent Boston Street waterfront warehouses, converted to residential condominiums with private docks for cruisers and sailboats. Then head north through several blocks of 1890–1910 vintage row-houses in superb repair. All have white marble front steps—all scrubbed, some with carpeting. Many are distinguished by painted scenes on the doorscreens. This

Row-houses with their famed white marble front steps in enduring ethnic neighborhoods on the Coastal Plain in eastern Baltimore . . .

. . . and their crowded, intensively used backyards.

is Highlandtown—a workers' district for generations. Signs along Eastern Avenue indicate the variety of ethnic neighborhoods—Polish, Czech, German. If it's near lunch time, ask for directions to Hausner's restaurant. If it's not lunch time, go to the high ground in Patterson Park and savor the panorama of row-house facades and rooftops, a rich variety of national parish church steeples, and the harbor.

Work your way northward past the big Johns Hopkins University Hospital—ranked as one of the country's finest—on Broadway. Follow Belair Road across the valley of Herring Run, with its wide strip of public park, and northeastward through two miles of middle-value, early twentieth-century neighborhoods to the Beltway (Interstate 695). If you missed lunch in Highlandtown, you might stop at one of the inviting neighborhood crab houses along Belair.

SUBURBS AND EXURBIA ON THE PIEDMONT

Move quickly northwest (but within legal speed limits, of course) on the Beltway. Like most of the country's circumferential freeways, this one generally separates landscapes built before 1960 and after 1960. The upper-value, auto-era residential development is typical of the Piedmont in suburban Baltimore County. In the post-World War II landscape the popularity of row-housing persists well into its third century.

Leave the freeway on Dulaney Valley Road (Route 146) and head south into central Towson. The campus of Goucher College (established for women in 1885) is on the left, and the extensive, modern campus of Towson State University (founded in 1866) lies about a mile to the southwest.

Before you reach central Towson, turn northwest onto York Road and pull into the small Prospect Hill burial ground on the right. Its ridgetop location affords tree-framed views of the Piedmont landscapes of suburban Baltimore County. A quiet, shady place to contemplate the grave markers from the rural early nineteenth century, it contrasts profoundly with today's heavy traffic, modern offices, shopping malls, apartments, and condominiums all around you.

Baltimore, Maryland, to Newark, Delaware

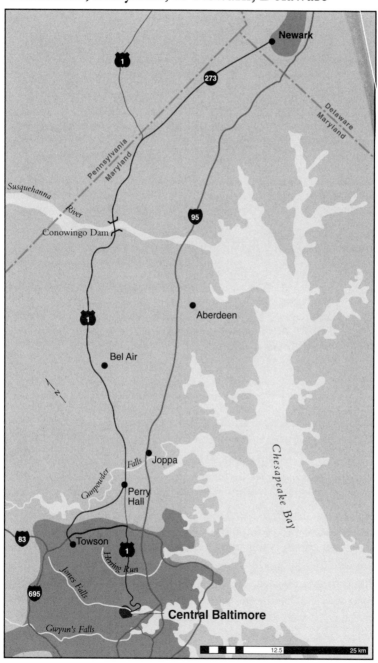

The automotive age has penetrated Baltimore housing styles in many places in the Piedmont suburbs of Baltimore County. Also note the young vegetation on the lots versus the established trees on old fence lines.

An important part of the modern surroundings reflects the efforts of Massachusetts contractor and former governor John Volpe—a major developer here in the boom years of the 1960s, when Spiro T. Agnew was Baltimore County administrator. Both became nationally visible in the 1970s as secretary of transportation and vice president, respectively, under President Richard Nixon.

Moving north from central Towson on Route 146, stop at historic eighteenth-century Hampton House, whose builder owned tobacco, timber, and iron lands extending twelve miles from Towson to Joppatown on the Chesapeake Bay. Continue north a short distance from Hampton House, then turn east on Seminary Avenue to Loch Raven Road. Then, turn onto Cub Hill Road to North Wind Road. On the former land of the Hampton estate, you now pass expensive subdivisions, the forested municipal watershed of

Gunpowder Falls, and exurban acreages above the Gunpowder River valley. The succession of byways returns to U.S. Route 1 at Perry Hall, another typical suburbanizing, former rural village of Megalopolis.

Continue northward through exurbia to the mile-long, hundred-foot-high Conowingo Dam, where the Susquehanna River crosses the Fall Line. For an impressive view of both the river and the dam, take the side road that winds down the wooded bluff to a riverside fishing facility near the power plant. Development of the turbine in 1910 made high dams on large rivers feasible, but the steam turbine and burgeoning market for electric power quickly shifted the bulk of the subsequent new generation to coal. Now, eighty miles upstream, the Three-Mile Island nuclear plant feeds power into the same regional electrical network. The search for long-term answers to ravenous energy needs continues.

East of the river, you leave the Metropolitan Statistical Area of Baltimore and enter the Consolidated Metropolitan Statistical Area of Philadelphia-Wilmington-Trenton. Continue east to Newark, Delaware, home of the state university (founded in 1833). This pleasant college town is a good place to seek overnight accommodations. Nearby alternatives are the restored colonial town of New Castle, on the shore of the Delaware River estuary, or downtown Wilmington.

Newark, Delaware, to Philadelphia, Pennsylvania

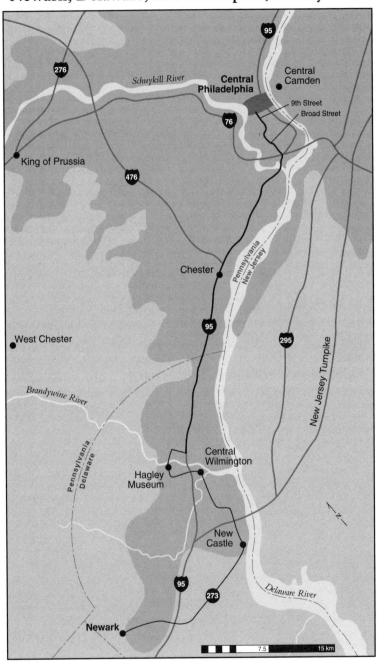

△ Days Three and Four

METROPOLITAN WILMINGTON AND PHILADELPHIA

Prologue

The U.S. Census Bureau lumps the three-state cluster of thirteen counties around Philadelphia, Pennsylvania; Wilmington, Delaware; and Trenton, New Jersey, into a Consolidated Metropolitan Statistical Area (CMSA). Our route winds through much of the Philadelphia and Wilmington portions of the CSMA but misses Trenton.

Total population of the thirteen counties rose substantially from 4.3 million in 1950 to 5.7 million in 1970, but only slightly thereafter to 5.9 million in 1990. Meanwhile the black component increased from about 9 percent in 1950 to 14 percent in 1990. Like all of Megalopolis, numbers in the old core declined and became much more heavily black, with offsetting increases in the predominantly white suburbs. In the central county of Philadelphia, total population declined from 2.1 million in 1950 to 1.6 million in 1990, while the black share rose a quarter-million, from 18 percent to 46 percent.

The Fall Line runs northeast-southwest through Trenton, Philadelphia, and Wilmington. Each city grew initially at a critical site on the tidal Delaware River estuary. The New York City-Washington, D.C., rail corridor, following the upper margin of the Coastal Plain, crosses the Delaware at the head of tidewater in Trenton, passes through all three cities, and links their historic port districts.

Settled by Swedish colonists in 1638, Wilmington was occupied by Dutch, then British. Quaker settlers led by William Penn (b. 1644; d. 1718) established Philadelphia in 1682. The three nations contested for possession of the Delaware River Valley in the mid-seventeenth century.

The central business district (CBD) of Wilmington developed adjacent to the original settlement and port at the mouth of the Christina River, a small tributary of the Delaware. Older, rail-industry sectors of the city developed on the Coastal Plain near the Christina River port and northeastward from the CBD. The high-value residential sector reaches northwestward on the Piedmont.

A major rail- and water-oriented industrial corridor follows the western shore of the Delaware estuary from Wilmington to Philadelphia. Colonial settlements along the shore swelled to blue-collar industrial satellite cities in the late nineteenth and early twentieth centuries. Automobile Era industries have filled voids that remained from the rail era, and a few industries have been added on the open land on the New Jersey shore, across the wide, deep estuary.

Philadelphia's CBD grew along the east-west Market Street axis running from the Delaware River to the Schuylkill River. The major rail-industry corridor crosses through the northeast sector and skirts north of the CBD and the oldest built-up area. A narrow but vital bundle of rail lines ran straight north from the CBD to that major corridor, then branched into multiple lines running northward toward the anthracite coal fields. On the west side of the Schuylkill, opposite the CBD, the Pennsylvania Railroad's routes branched southwestward to Washington, D.C., and west to Pittsburgh, Pennsylvania, and the Middle West. The latter route was the historic "Main Line." A large industrial satellite grew at Camden on the New Jersey side of the Delaware River, and was joined to the central city by ferries and to New York City by rail lines.

Railroad Era residential development was mainly within Philadelphia County (now also the city), although important early suburbs strung along the rail routes—especially the Main Line. Highest-value residential areas were principally in the northwest sector and the Main Line suburbs. Today's predominantly black neighborhoods are mainly in the railroad-era central city and the old cores of the industrial satellites.

Automobile Era development has spread far in all directions—over the Piedmont in a broad arc across Chester and Montgomery counties (southwest, west, and north of the city) and on the Coastal Plain, especially in sprawling suburbs all around Camden. Accessibility to the major, somewhat more compact growth areas on the Piedmont is provided by the fairly dense legacy of suburban rail lines, along with a rather limited freeway web. In contrast, the sprawling expansion on the New Jersey side has been facilitated by most of the area's circumferential freeway mileage, radials from four Automobile Era bridges across the Delaware, and just one post-World War II suburban rail line. Government and private investors have accomplished important redevelopment and renovation in the CBD.

The largest of the family of suburban retail-office developments adjoins the first and busiest freeway interchange in the Pennsylvania suburbs, at King of Prussia. Another is near major interchanges at Cherry Hill, New Jersey.

Public open spaces are mainly on the Piedmont. The most distinctive open space in Megalopolis opens from the western suburbs into the farming area of Lancaster County. The eastern fringe of the CSMA also includes a large amount of open woodland, scrub, and some agricultural land on the Coastal Plain

Guideposts along the Route

The road from Newark to New Castle cuts through a classic piecemeal landscape of farm abandonment, clustered and scattered single-family homes, condominiums and apartments, strip malls, and small industries. The area illustrates the interplay of utility extensions, land development, roads, and land-use planning on the frontiers of dense urbanization in Megalopolis.

HISTORIC SETTLEMENTS AT THE FALL LINE

After a brief transect through the turn-of-the-century blue-collar parts of the city, enter old New Castle, Delaware. Established by

Swedes in 1651, then taken for the Dutch by Peter Stuyvesant (b. ca. 1610; d. 1672), and then by William Penn in 1682, the place changed from a modestly important colonial port to a blue-collar residential backwater, but more recently to a mint collection of restored eighteenth-century landscapes. As you roam the quiet square and streets, read the plaques, and you'll be intrigued by the buildings and procession of notables who once used them.

Glimpse the remains of the port on the Christina River as you head north into downtown Wilmington—dominated by the office-bank-hotel complex of the DuPont chemical corporation headquarters. Work your way northwest through gentrified neighborhoods and the well-preserved residential sector on the Piedmont.

In northwest Wilmington, get directions to the Hagley Museum and Library (telephone 302-658-2400), hidden at the bottom of the Brandywine River gorge below Mill Road. Here at the Fall Line, E. I. Du Pont de Nemours (b. 1771; d. 1834) built his gunpowder

Restored colonial-era homes around the central square at the historic port of Newcastle, Delaware.

mill and his home in 1802. A student of French chemist Antoine-Laurent Lavoisier, he came to the United States from France in 1800 with his father, Pierre-Samuel Du Pont (b. 1739; d. 1817). The elder Du Pont is remembered as a transatlantic scholar and statesman. He played an important role in the early phases of the French Revolution, urged his government to sell Louisiana to the United States, and drafted a national education plan for President Thomas Jefferson. You'll recall a plaque on a building at New Castle in which the Du Pont family entertained the Marquis de Lafayette on his return to the United States in the 1820s. These landscapes provide still more reminders of the location of Megalopolis at the hinge of the northwestern Europe-U.S. corridor.

Resume the journey east on New Bridge Road past the DuPont corporation's vast research campus and back to Interstate 95. A half-mile beyond the interchange, down on the Coastal Plain, in a small minority-owned neighborhood grocery, a prominent sign tells you that food stamps are accepted and you can buy a copy of the radical tabloid, *Workers World*. It's one of the many vignettes reminding you, again and again, that great social distances go with short physical distances in Megalopolis.

TRANSPORTATION CORRIDOR ALONG THE DELAWARE RIVER

Northeast of Wilmington, Interstate 95 parallels the rail-industry-shipping corridor along the Delaware River—long a major concentration of heavy industry in Megalopolis. Helicopter, railroad car, shipbuilding, and paper industries line the route northeast of Chester, Pennsylvania.

The twelve-hundred-acre Tinicum National Environmental Center lies between the industrial strip and the river about five miles beyond Chester. The area is a remnant of the tidal marshes that once lined the river. A haven for bird wildlife, the preserve is also an important urban environmental preserve and research area. Like most natural preserves in Megalopolis, it is a product of the Automobile Era. Widespread auto ownership brought the urban invasion of the countryside, visible disturbance of pristine landscapes, and awareness of the problem among the mass of urban citizens.

Dikes, built by seventeenth-century Dutch settlers, remain. The early dikes are trivial in comparison with twentieth-century modifications of the marshes. For about the next five miles after Tinicum, made or drained land underlies the Philadelphia International Airport, extensive refineries and petrochemical plants, a part of the Philadelphia Naval Shipyard, and the extensive park lands around the professional-sports arena and stadiums for football, basketball, and ice hockey. The shipyard is in part a legacy from the eighteenth century. But the other developments belong to the early Automobile Era, when land near the city centers became scarce but was still much desired.

In the Automobile Era, it became common to drain and fill land, previously deemed unbuildable, to accommodate extensive, large-scale developments as near as possible to the established city centers. That kind of land-making was especially common before the economic weakening of the central cities, and before the rise of environmentalism. Like so many geographical features, the made-land areas are unique products of time, site, and location.

INTO CENTRAL PHILADELPHIA

Northward on Broad Street and northeastward on Passyunk Avenue the route plunges into the nineteenth-century neighborhoods of South Philadelphia. Block after block, traditional neighborhood-oriented shops front the avenue, and row-houses wall the side streets. Follow 9th Street north from Passyunk Avenue through the heart of the Italian Market. On a busy summer day continuous rows of sidewalk stands supplement open storefronts and crowd the sides of the narrow street. Small, cosmopolitan swarms of humanity press through the narrow space on the walks and mingle with slow-moving vehicles on the street. Colorful displays of goods and food adorn walls, windows, and curbside tables. The sights, sounds, and smells resemble the Old World. You'll be tempted to leave your car on a side street and take a brief stroll.

At South Street turn east from 9th Street to 5th Street and continue north on 5th into central Philadelphia—the corridor from the Delaware River to the Schuylkill River, specified in the 1682 plan. South

The lively scene in South Philadelphia's Italian Market.

Street marked the south edge of the corridor, Vine Street the north edge, Market Street the east-west axis, and Broad Street the north-south axis. City Hall Square was the planned center.

Ethnic neighborhoods give way to gentrification as you cross South Street into the Society Hill district. The area was originally a land grant from William Penn to the Society for Free Trade—a group interested in Pennsylvania economic development. It is now a showplace of central-city renewal. Strong revival of the district's name accompanied extensive, high-priced, high-rise and row-house redevelopment as well as rehabilitation of colonial buildings since the 1960s.

LIVING MUSEUM

Continue north on 5th Street through the Independence National Historic Park. Leave your car in one of the parking facilities beneath Independence Mall. If you're on schedule, it's midday.

The Bourse Building, facing Independence Square on 5th Street, was restored and converted in the 1980s, and its fine atrium offers an interesting variety of places for lunch. The building was the home of the Philadelphia Stock Exchange from 1895 until its demise in the Great Depression during the 1930s.

If you have not had a previous opportunity, allow a couple of hours to view the beautifully maintained shrines and landmarks in and near Independence Square. Beyond Independence Hall and the Liberty Bell, a half-mile stroll takes you to Carpenters' Hall, Congress Hall, and the first and second National Banks of the United States.

Other historic buildings include the largest U.S. mint and a remarkable cluster of specialized museums endowed by past civic leaders and organizations. One, the Atwater Kent Museum (its industrialist-founder's name is familiar to everyone who owned a radio in the early days of American broadcasting) has exceptional displays on aspects of the historical development of Philadelphia from 1680 to 1880.

If you have visited Independence Square before and have some spare time, ride one of the country's few remaining nontourist, city streetcar lines. The Germantown Avenue cars head north on 11th Street and run for miles through predominantly black neighborhoods of North Philadelphia and Germantown (settled by German religious minorities in 1693), the middle-class racial mixing zones of Mount Airy, and virtually all-white Chestnut Hill. The route traverses nineteenth-century middle- and lower-value neighborhoods, then fashionable late nineteenth- and early twentieth-century streetcar suburbs.

The traverse through North Philadelphia provides at least a small close-up of everyday life amid a remarkable mélange of crowds and isolation, maintenance and ruins, purchasing power and poverty, style and squalor. Through Mount Airy, the landscape reflects the profound commitment and effort of households and institutions on the frontier of racial integration.

Census statistics support impressions we derive from the landscape. In the past two decades the white population of Philadelphia has fallen by 850,000, while the black population has risen by

Age, contrast, and continuity on Independence Square, Philadelphia.

Looking eastward on Market Street, from Independence Mall past the facades of the old Lit Brothers department store to City Hall and Penn Center, Philadelphia.

350,000. The net loss has been half a million. In turn, the German-town trolley ride puts life into the statistics. In its way, the route is an important national historical symbol, stemming—like the enshrinement of Independence Square itself—from "the course of human events."

Following the midday walking tour or the alternative street-car trip, proceed from Independence Square west and north through the central business district. Its diverse facades convey an accurate impression of early economic power, continuity, aging, and a partly fulfilled, long-term program for renewal and rebuilding.

Towers of major financial institutions frame a westward view of City Hall, capped by a giant statue of William Penn. The office towers reflect legacies from Philadelphia's early commercial and

*Gentrification of a bit of the cramped colonial city, near the Schuylkill
River on the edge of downtown Philadelphia.*

industrial leadership. With its combination of openness to many different ideological and nationality groups, powerful business and social elite, and proximity to the richest colonial hinterland east of the northern Appalachians, it was an unusually energetic place. It was the leading commercial and industrial center and largest city in the United States from colonial times until 1820. Momentum from those times was a major force in subsequent growth, at least through the Railroad Era.

Among notable buildings on Market Street are the nine-teenth-century Lit Brothers department store (now a bank build-ing), the 1970s Gallery (one of the earliest major downtown mall developments, with department stores, more than 100 shops, and the modern downtown terminal for subway trains and the majority of suburban rail lines), the famous, monumen-tal John Wanamaker department store, and (west of City Hall) the two-block Penn Center, bellwether of the post-World War II downtown redevelopment scheme under the direction of major urban designer, Edward Bacon. The massive, nineteenth-cen-tury Reading Terminal faces Market Street, with a famous food and merchandise market hidden beneath its cavernous train shed a block to the north. The new convention center is rising north of the train shed, on the site of the former yards. Just east of the old terminal area, Arch Street passes the edge of tradi-tional Chinatown.

The Reading Railroad, a major anthracite carrier, also initiated the extensive suburban service still running north and northwest of the city. The terminal for western and southwestern suburban rail service, long operated by the Pennsylvania Railroad, adjoins the retail-office-hotel towers of Penn Center. The Pennsylvania Rail-road, built under Philadelphia leadership, was known at the turn of the century as the "standard railroad of the world." Conrail, still headquartered in Penn Center, is its corporate descendant.

PIEDMONT PARKS AND MAIN LINE SUBURBS

From the Philadelphia Museum of Art at the northwest edge of downtown, Fairmount Park extends more than seven miles into the Piedmont, up the Schuylkill River and tributary Wissahickon Creek. The campus of the University of Pennsylvania is downriver, across the Schuylkill. Distinctive among Ivy League schools for its early emphasis on science and modern languages in addition to the classics, "Penn" (*not* Penn State) emerged from a union of early predecessors as the College of Philadelphia in 1755.

Following Kelly (East River) Drive through the park, the impression of early economic power, continuity, and aging continues. Pass the famous Boat House Row, home of old and elite rowing societies. Then, among the trees across the river, glimpse the steel-and-glass Memorial Hall, from the 1876 World's Fair—a great showing of U.S. industrial and commercial progress in the nation's first century. Next pass Robin Hood Dell, summer home of the Philadelphia Symphony.

At City Line Road, climb out of the gorge, cross the Schuylkill River, and enter the Main Line suburbs. Wind westward on Montgomery Avenue through Bala-Cynwyd and Narberth, past a pioneer suburban shopping center in Ardmore, and westward through Haverford, Bryn Mawr, and Villanova. Each of the latter three Main Line suburbs is home to a distinguished college of the same name (founding dates 1833, 1885, and 1842, respectively). The newer and most affluent areas are farther from the rail line. Keep your detailed map in hand and detour northward toward Route 23. Explore some of the side streets that curve among the Piedmont ridges and glens. The scene leaves no doubt that this is one of the most extensive ostentatious residential districts in the country.

Cross the Schuylkill River on Interstate 476 and circle through Conshohocken, typical of the old industrial satellites along the canal and the early railroad to Reading and the anthracite fields. Norristown is another, larger city of the same genera. Both are now engulfed in the metropolitan urbanized area. I-476, only recently completed, is the metropolitan area's long-gestating western circumferential freeway.

Return to the Schuylkill Expressway (Interstate 76) and continue westward to the complex interchange area at King of Prussia. At the first major freeway interchange in the metropolitan area,

Central Philadelphia, to King of Prussia, to Levittown, Pennsylvania

An impressive single-family home in the auto-era portion of the Main Line suburbs.

this was the location of a prototype large-scale, planned office-industrial park in the post-war era. A quarter-century of subsequent development has produced an exceptionally large concentration of retail, office, and hotel facilities.

If you are on schedule, you'll reach the King of Prussia area in the late afternoon of the third day. This is a convenient location to spend the third night of the excursion. But don't linger here after you check in at your hotel or motel. There's still time for a pre-dinner drive into the Amish and Mennonite farming country of Lancaster County.

TRANSITION TO RURAL LAND AND LEGACIES

By coincidence, this massive 1960s-to-1980s suburban center adjoins the spacious national park embracing the site of the 1777–78 winter encampment of George Washington's troops, at Valley Forge, during the British occupation of Philadelphia. When you leave your hotel or motel for the dinner expedition, find State Route 23 (Valley Forge Road) and head west through the Valley Forge National Historical Park.

Continue west on Route 23 through exurban areas of large-lot subdivisions, restored farmhouses, horse farms, and some working farms. The crossroad at Warwick leads to Hopewell Village, a National Park Service restoration of one of many pre-1830 iron-smelting communities based on charcoal and local Piedmont ores. Around Morgantown, the Pennsylvania Turnpike (I-76) interchange provides easy access to Lancaster County (population 423,000) from metropolitan Philadelphia and for weekend traffic from New York City. Hence a gaggle of tourist-oriented outlet stores, gift shops, and fast-food establishments is clustered here.

Beyond Morgantown, the landscape along Route 23 changes dramatically to lush rolling crop and pasture lands and big, handsomely maintained farmsteads. These are the Amish and Mennonite settlements of Lancaster County—a legacy from the initial settlement at Germantown. Today the area is a cultural island, where combustion engines, electricity, and telephones are shunned in deference to abiding beliefs from sixteenth-century Protestant Europe. The combination of its land, people, and market location makes it one of the richest agricultural counties in the country. Today's diversified agricultural production includes corn, wheat, barley, and a variety of specialty and horticultural crops (including tobacco, a big cash crop for the Amish), dairy products, beef cattle, hogs, and poultry. Dairy farming is still the principal agricultural activity of the Amish.

You have also entered the Conestoga Valley, where early German craftsmen built the Conestoga wagons made famous in the lore of western settlement. Today's manufacturing industries, many of them farm-oriented, operate on the edges of the small towns. Lancaster (population 56,000) is an important marketing, service, and industrial center, as well as the home of esteemed Franklin and Marshall College (founded in 1787).

The highway passes several Mennonite churches and buggy sales and repair shops. As you approach Route 897, inquire about the location of the Shady Maple Diner—a modestly priced restaurant with local fare and patronage. There are other comfortable restaurants in the hamlet of Blue Ball or in the larger farm-service center of New Holland, a few miles farther down Route 23.

If some daylight remains when you return to King of Prussia, again with map in hand, explore the extensive retail malls and office-industrial parks. Retail volume and office space make this a second metropolitan downtown, but with a vastly different density and ambience from the first one in central Philadelphia. You'll be tempted to speculate on the comparative function and appearance of both in the year 2100, when this one is as old as the first one was at the time its modern renewal got under way.

Begin the fourth day with a dash (east) on the Pennsylvania Turnpike (I-76) from King of Prussia to Levittown. Built originally through largely open country, the Turnpike now cuts across the grain of suburban settlement—lines of older centers along five radial commuter rail lines and countless newer subdivisions between the radials. Although the great bulk of development is auto-oriented, there is a paucity of radial freeways.

BACK TO THE COASTAL PLAIN

Leave the standard freeway landscape design of the Turnpike, and plunge into an archetypical blighted commercial arterial strip of early post-World War II vintage along Route 13. Then turn into Levittown Boulevard and enter a prototypical post-war, comprehensively planned tract-housing development.

The New York City-based Levitt brothers were visionary entrepreneurs who sought to do something significant, quickly and profitably, about the unprecedented housing shortage in the late 1940s. The crisis stemmed from a fifteen-year hiatus in home building through the Great Depression and World War II, followed by the deluge of war-delayed new families.

This Levittown followed the Levitt brothers' initial development on Long Island. Like that one, it combined a metropolitan fringe location with easily prepared, flat, sandy land, innovative assembly-line construction, and thoughtful layout of streets and land uses. Nearly half a century later, the place is a museum of diverse individual modifications of what were originally standardized cottages. Meanwhile, the community has maintained its integrity and appearance remarkably well. It has done so despite the virtual collapse of its shopping center in the face of competition from newer regional malls, growth trends away from this part of the metropolitan region toward the Piedmont, and the near-extinction of the major local employer, the nearby Fairless Steel Works. (Note: Please refer to the map at the beginning of the next chapter for directions to Atlantic City from Levittown.)

The bridge from Bristol to Burlington, New Jersey, provides a panorama of the upper reaches of the Delaware estuary and the Coastal Plain. The Fairless Works of USX (formerly United States Steel Corporation) stands at the head of modern ocean navigation a few miles upstream. Built after World War II, the plant combined ore from South America and Africa with Appalachian coal to supply the eastern steel market. Long one of the largest old-technology mills in the country, it appears to be nearing the end of its life. The industrial and state capital city of Trenton, New Jersey, lies fifteen miles upstream, just beyond which lie the outer suburban reaches of New York City.

At the New Jersey end of the Bristol-Burlington Bridge, begin the seventy-mile run across the flat Coastal Plain to Atlantic City. Follow Route 541 from Burlington to Interstate 295, south to State Route 73, and on to the Atlantic City Expressway. The route passes through the eastern margins of the Philadelphia urbanized area, the southern margin of the Pinelands, and the northern margin of the agricultural area of southwestern New Jersey. Over the first half of the route, scattered, modest residential subdivisions or isolated cottages and scrub woodlands or brush dominate the flattish plain.

Occasionally you'll glimpse a cultivated field or an orchard. Cropland occupies about one-tenth of the land in this area. The western and southwestern counties of New Jersey

Suburban development expanded into the Coastal Plain farmlands of Camden, New Jersey, during the 1960s.

have been major producers of vegetables and horticultural produce for the markets of Megalopolis. Canning and freezing were pioneered in the area. The Campbell Soup Company began in Camden. Seabrook Farms, a few miles to the south, was once one of the largest vegetable-producing operations in the world. Dairying and poultry have also been important. But intensity of farming has declined in the past quarter-century.

This area has been affected by competition from the tropics and subtropics, rising land values accompanying urbanization, and comparatively high labor costs. Crop acreage is down in most of the area, and the importance of poultry and dairying has diminished sharply. Inflation-adjusted income from agriculture is down by one-quarter or more. By comparison, during the same period, farm income doubled in Lancaster County, Pennsylvania, and increased by one-third in the vegetable-producing districts of southern Florida.

You leave the census Consolidated Metropolitan Statistical Area of Philadelphia where the route enters the Atlantic City Expressway near Hammonton, halfway between Camden and Atlantic City.

△ Day Four

THE NEW JERSEY SHORE

Prologue

For the last twenty miles on the mainland, the Atlantic City Expressway crosses Atlantic County. The county defines the Atlantic City Metropolitan Statistical Area (MSA). Crowded between the Atlantic Ocean and the Absecon Bay, the Railroad Era city's population dropped from 61,000 in 1950 to 35,000 in 1990, while the black proportion has grown from one-fourth to nearly 70 percent. Meanwhile, growth on the mainland has doubled the county population to nearly a quarter million.

Guideposts along the Route

TO ATLANTIC CITY

The mainland landscape along the Expressway is a mixture of woodlands and patchy residential development spilling out of Atlantic City. The sandy soil—though droughty and infertile—provides the groundwater catchment area for urban developments to the east. The Pinelands—a region of scrub pine and oak—extend thirty to forty miles north and south of the route. Although settlement here goes back to the 1700s, it was always sparse and marginal because of the poor soil.

Soon a skyline of towering hotels and condominiums marks the edge of the beach ahead. Cross Absecon Bay, enter the city, and

Levittown, Pennsylvania, to Raritan Bay, New Jersey

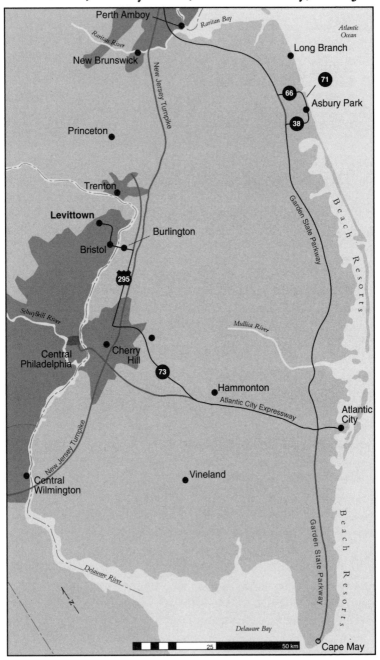

A glimpse of the changing landscape of Atlantic City: newly-cleared land for redevelopment amidst blocks of derelict railroad-era apartments, and 1980s casino-hotel on Absecon Bay.

then drive the last mile through virtually treeless blocks of houses, flats, and stores. Many structures are dilapidated, most are fully used, but a substantial number are more or less abandoned. Extensive areas are in various stages of clearance and redevelopment. This was the high-density city for service workers and bargain-seeking visitors in the Railroad Era—walking or streetcar distance from both the Boardwalk and the depot.

Park in one of the ample facilities within a block or two of the Boardwalk. If you're on schedule, it's lunchtime, and here you are presented with multifarious choices. A possibility is one of the cafeterias or restaurants associated with the casinos. If you're not inclined to gamble, you may contemplate the instincts of the crowds who do.

The wide Boardwalk separates the less crowded beach from the more crowded buildings—the brick towers of the Railroad

Era hotels, taller glass-and-plastic towers of the post-1970 casino hotels, and one-story souvenir and food shops. The crowd around you is a cross-section of Megalopolis. They have come in buses, limousines, and classic cars. Senior citizens, middle-aged couples with children, youths, entertainers, card-dealers, and showgirls wear what seems like the full range of costumes available in the world. When you have observed everyone else, mirrored walls everywhere enable you to see yourself from every possible angle.

The city boomed initially as a product of the railroad connections and expansion of buying power in the latter half of the nineteenth century. Its massive hotels and "Steel Pier" attracted major conventions, and the city billed itself as "The World's Playground." It gradually aged into a rather threadbare attraction on the low-price end of the Megalopolitan market in the Automobile Era, then revived with the legalization of casino gambling in the 1970s. Per capita retail and service sales, adjusted for inflation, have doubled since the 1960s. After a long lapse, rail passenger service has been restored. The twelve casinos grossed about $3 billion in 1990.

When you return to your vehicle, drive north (Route 87) over the bridge and circle through neighboring Brigantine. Near the beach, a small 1920s-vintage high-rise hotel and bungalows of the same age mark the extent of the place before World War II. With no rail connections, it awaited the Automobile Era for settlement. Today you drive through several square miles of post-1970 housing tracts and condominiums, with accompanying strip malls and marinas, and more to come.

If Atlantic City is unique, Brigantine's development is much more representative of vast stretches of beach that had been either inaccessible or only partly developed until the Automobile Era.

BACK TO THE "REAL WORLD"

Return to Atlantic City and follow U.S. Route 30 back (west) to the mainland and the Garden State Parkway. After about seven miles, at the Mullica River estuary crossing, you glimpse some remaining fisheries and an early waterway to small settlements in the Pinelands. You also enter Ocean County. So take a deep breath—

Looking east toward the Atlantic Ocean beach through crowded layers of vacation cottages in Seaside Heights, New Jersey.

you have now entered the New York City Consolidated Metropolitan Statistical Area.

Farther north on the Garden State Parkway, interchanges at Route 72 (to Ship Bottom) and Route 37 (to Seaside Heights) provide access to long rows of closely spaced, small oceanside towns. Rail lines from New York City ran down those sandbars at the turn of the twentieth century, and the towns were the stations. Clusters of closely packed little cottages tend to mark the old station districts. Automobile Era homes fill the interstices. The state Transit Authority has restored commuter rail service from the beaches between Long Branch and Bay Head (east of the exit to Route 70) to Midtown Manhattan. Meanwhile, south of Seaside Heights, ten-mile-long Island Beach State Park has been salvaged from the jaws of development.

Leave the Garden State Parkway (exit 98) to go out to the coast at Asbury Park-Ocean Grove on Route 138. Founded with strong

Luxurious vacation residences preserved from the Railroad Era, Asbury Park, New Jersey.

religious influence in 1869, the city residential core is a museum of Victorian resort development for the elite and would-be elite. This detour off the tollway also brings you near enough to the bays to observe a few samples of the multi-billion-dollar fleet of private recreational boats that uses the shores of Megalopolis. The fleet is easy to observe, because almost all of it is resting in port most of the time.

Proceed north through Deal's luxury residential areas and return to the tollway on Route 36. The first view of the Manhattan skyline, twenty-five miles across New York's Lower Bay, comes as a thrilling surprise along a low ridge about fourteen miles northwest of your re-entry to the Garden State Parkway. This view marks your symbolic entry to the heart of metropolitan New York.

△ Days Four, Five, and Six

METROPOLITAN NEW YORK

Prologue

New York City accounts for nearly half the population of Megalopolis. Its census-defined Consolidated Metropolitan Statistical Area (CMSA) includes twenty-four counties in three states. The City of New York comprises five of those counties—the boroughs of Manhattan (New York County), Brooklyn (Kings), Bronx, Queens, and Staten Island (Richmond).

The metropolis is massive, crowded, expensive, and a complex reservoir of wealth and poverty. In 1990 the central-city population was 8.5 million, the remainder of the CMSA 9.5 million, for a total of 18 million. An average square mile—the area of eight average farms back in Lancaster County, Pennsylvania—has 67,000 residents in Manhattan, 33,000 in Brooklyn, 24,000 in the Bronx. The twenty-three-county area has 7.5 percent of the nation's metropolitan population, 15 percent of the metropolitan local-government revenue and expenditures, 11.1 percent the metropolitan personal income, and 12 percent of the metropolitan poor people (see map on p. 44).

The area's population growth has been stagnant since the 1970s. Population of the CMSA rose from 14 million in 1950 to 18.1 million in 1970, continuing a 300-year-long growth trend. But it dropped slightly to 18 million in the twenty years after 1970.

Meanwhile, New York City's part of that total was 7.9 million in 1950 and 1970, and fell back to 7.3 million in 1990. But beneath the apparent stagnation, there have been massive changes in the past two decades. While New York City's white population declined by 2.4 million, its black population increased by 1.8 million, and there were commensurate increases in numbers of Hispanics (especially Puerto Ricans) and Asians (especially Chinese). Since the metropolitan population was almost unchanged, the central city loss had to be compensated by equal suburban gain. Hence the suburbs continued to boom despite regional stagnation. Hence, also, despite rapid black population growth, thousands of structures were abandoned in the central city.

LAND AND DEVELOPMENT

The physical setting of the New York metropolis is varied and naturally attractive, though much of it is all but buried under one of the world's largest masses of urban structures. It overlaps the nexus of the Appalachians, Coastal Plain, and Ice Age glaciation.

The elongate north-south upland blocks and trenches of southwestern Connecticut, the Bronx, Westchester County (New York), and the Hudson River Valley and Palisades are part of the crystalline-rock Appalachian uplands. So are the glacial ponds and mountain ridges of the northern New Jersey suburbs and the Watchung Mountains and hilly Piedmont in the suburbs to the west and southwest.

The bluffs and hummocks of northern Long Island and the sandy plains, coastal marshes, and beaches of southern Long Island are on the glaciated Coastal Plain. Familiar names such as Flatbush and Forest Hills reflect the way the land looked to early settlers and developers. Meanwhile, on the mainland south and southwest of Newark, the flattish lowlands, estuaries, and beaches of the New Jersey suburbs are on the unglaciated Coastal Plain. Land typical of all of these varied regions comes together in Manhattan and Staten islands.

Development began with the Dutch fortification and port of New Amsterdam at the lower tip of Manhattan in 1624. Ports on the New Jersey mainland were established on the Passaic River estuary at Newark in 1666 and the Raritan River estuary of New Brunswick in 1681. When Alexander Hamilton (b. 1755; d. 1804) established the industrial town of Paterson at the Passaic Falls, in 1791, New York City's population had grown to nearly 40,000. New York vaulted into national population and commercial pre-eminence with the completion of the Erie Canal in 1825.

In the early steamship and railroad years, from about 1830 to 1870, expansion of docks focused on the established lower Manhattan shores, in Brooklyn, and on the opposite side of the Hudson River at the terminals of rail lines from the interior in Jersey City and Hoboken. Railroad-car ferries linked the three major dock areas. Suburban train service to the lower Manhattan business core terminated at the Jersey docks and in Brooklyn, and ferries linked both to lower Manhattan.

The second half-century of the Railroad Era, from the 1870s to the 1920s, was a time of great growth and metropolitanization. Opening the era, the Brooklyn Bridge (built between 1870 and 1873) revolutionized the link to lower Manhattan. Freight traffic multiplied, and several miles of new docks were added on both sides of Manhattan north to Midtown, along the Brooklyn water-front, and north from Hoboken. Suburban train service extended to the edges of today's twenty-three-county metropolitan area. The five boroughs were unified to form New York City in 1898.

Then, early in the twentieth century, electrification facilitated two projects that changed the face of the city. The New York Central Railroad put its tracks into a tunnel from upper Manhattan south to 42nd Street. In the greatest privately financed engineering project in history, the Pennsylvania Railroad used new technology to tunnel from Hoboken to the southwest corner of Queens— beneath the Hudson River, Manhattan Island at 32rd Street, and the East River. The same project built the 3½-mile-long(!) Hell Gate Railroad Bridge (built in 1914 and 1915) across the western end of Long Island Sound, from Queens to the Bronx. With a new line from Hell Gate to the tunnel, trains could run to and through Manhattan directly from New England, the South, and West.

The new Grand Central Station and Pennsylvania Station became the gateways to the suburbs and to the rest of the United

States. A new center for New York City emerged as Midtown, between 30th Street and Central Park. Land values and investment patterns shifted the city's major retail, hotel, entertainment, and general office activity to the area. Lower Manhattan gradually stabilized as the residual concentration of finance—which, of course, was itself a major growth sector of the economy. The intervening built-up area gradually languished, with a changing variety of specialized commercial, light-industrial, and residential districts. And the leading edge of office, hotel, and shopping development pushed northward around the southern part of Central Park.

Meanwhile, tentacles of rail-industry land use reached across western Brooklyn and Queens and the New Jersey Coastal Plain. The subway system and feeder streetcar lines extended the dense, pre-auto residential and neighborhood commercial development essentially to the limits of New York City, except for remote central and southern Staten Island. Even Long Island's south shore beach was intensively developed at Coney Island. On the New Jersey side, lack of bridges and subways made commuting to Manhattan less important, while most of the metropolitan area's rail-oriented industry created many local jobs there. So the trans-Hudson side of the metropolis evolved as a quite different world from the New York side.

At the end of the Railroad Era, in the 1920s, metropolitan population had passed 8 million. Only half a century later it was 18 million. A new Automobile Era city of 10 million had surrounded the Railroad Era core and filled the interstices between the radial bead-strings of railroad suburbs.

The new city formed a huge 2,000-square-mile halo, with a circumference of more than 100 miles, surrounding the 500-square-mile pre-auto core. Compared with the core, the outer metropolis was far more extensive, much lower-density, predominantly post-World War II vintage, and set mainly on rolling uplands.

Because of physical barriers and sheer size, the Automobile Era city evolved as a collection of big, separate realms—northern New Jersey; the peninsula between the Hudson River and Long Island Sound; Long Island; and New Jersey south of Lower New York Bay. Several hundred thousand commuters continued ties to Manhattan, and several major bridges and tunnels had been built. But the great and growing majority tended to work and live within their separate realms. Decentralization and fragmentation have prevailed. Meanwhile, Manhattan's office and hotel concentrations continued to grow with the global economy. But, with some notable exceptions, the rest of the Railroad Era core suffered from aging and obsolescence, neglect, and massive turnover to lower-income residents and businesses. Investment in most aspects of public improvements had fallen far behind the trends in population growth and turnover. The commuter rail and subway lifelines of lower and Midtown Manhattan were in shambles. Traffic congestion had become serious everywhere.

During the 1970s and 1980s, some changes began to appear. Circumferential freeway rings were finally established. Billions of dollars were committed to renovate and rationalize the suburban rail systems, sewage and solid waste management, and power and communication systems. More breadth and diversity of approaches to maintenance and replacement of buildings and open space became apparent. The generational wave of the "baby boom" and the migrational wave from the rural South had subsided. Hence, the cultural conflicts and social disorganization that resulted from those events could gradually work themselves out through continuing social change, if commitments to the task were sustained. The net effect of these complex trends has been to make the landscapes of the New York region less dense but even more mixed and chaotic.

An impressively large array of public open spaces have preserved the diversity of physical landscapes, although many of these parks suffer from heavy, abusive use and neglect of maintenance.

More than 9 million people work in the region, 2 million on Manhattan Island alone. Despite the shift to other places, there are 1.5 million manufacturing jobs. One million work in retailing and wholesaling; 800,000 in finance and commerce; and 4.5 million in a multitude of government, business, and personal services—more than a million of those self-employed.

Raritan Bay, New Jersey, to Midtown Manhattan

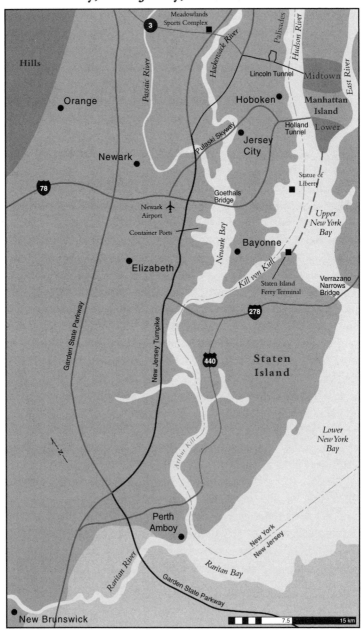

Guideposts along the Route

THE NEW JERSEY WORKSHOP

The route enters the heart of the New York metropolitan region where the Garden State Parkway crosses Raritan Bay, beside the industrial-railroad-port satellite city of Perth Amboy, and turns north on Interstate 95, the New Jersey Turnpike. For the next twenty-five miles, you cut through developments facing Manhattan from the New Jersey shore.

You quickly pass miles of rail yards—many derelict or cleared—and miles of Automobile Era chemical plants, refineries, tank farms, and salvage yards along the Arthur Kill. (Note: The Arthur Kill is the channel between New Jersey and Staten Island, New York, that connects Raritan Bay with Newark Bay to the north.) This is all part of the colossal, largely ignored enterprise needed to supply the materials and process the waste for Megalopolis.

The low, wooded ridgeline of Staten Island rises on the east, across the Arthur Kill. The skyline of Elizabeth, New Jersey, is visible on the left—just one of many such local clusters scattered within the solidly built-up area from here to New Haven, Connecticut.

Newark International Airport and three square miles of container docks, on made land along Newark Bay, sprawl on either side of the New Jersey Turnpike north of Elizabeth. Along with Kennedy and La Guardia airports on Long Island, Newark is one of the major airports that provide service in the New York hub— the country's largest air-transport hub, with nearly 40 million passenger boardings annually.

The Newark container port is by far the largest on the East Coast. Its volume is level, or slipping somewhat, because of competition from Norfolk, Virginia, and Jacksonville, Florida. Los Angeles-Long Beach in California surpassed Newark in the mid-1980s. The comparison with Los Angeles reflects a combination of the growth of Pacific sea lanes in the global system, the growing importance of the West and South in the U.S. system, and the improvements in technology that enable combined ocean-rail container shipments from Asia

A super-container ship outbound from Port Newark, New Jersey, passes between Staten Island and Lower Manhattan.

to reach the east-central states more cheaply through California than through New York City or Baltimore.

Next, the Passaic River enters Newark Bay from the northwest. Downtown Newark's skyline rises two miles upstream, at the river's historic head of navigation. Paterson is ten miles further upstream, at the falls. The Watchung Mountains form the western horizon. A massive, mostly Automobile Era suburban population of two million spreads between the two central cities and up over the Watchungs.

North of the Passaic River, the route follows the wide marshes and meadows along the Hackensack River. With the completion of the freeway network, this bypassed land suddenly acquired a high degree of metropolitan accessibility. One result was the Meadowlands Sports Complex, looming ahead on drained and filled land, home to horse racing and New York's two professional football teams.

Midtown Manhattan from the Palisades, 1980s. The Empire State Building (far right) still dominates the skyline. Forty- to sixty-story boxes represent post-World War II office booms. The small low facade behind the docks represents turn-of-the-twentieth-century architecture. Big passenger docks once served transatlantic liners, occupied at this time by cruise ships that had not yet shifted to the port of Miami in Florida. Rotting pilings of abandoned rail-freight docks and cleared, former rail-yards in the New Jersey foreground await gradual redevelopment. A recreational sailboat represents a growing majority of the Hudson River traffic.

Leave the Turnpike at the exit for Route 3 and the Lincoln Tunnel, and go east into the deep cut through the upland of the Palisades lava block that separates the Hackensack meadows from the Hudson River. After about 1.7 miles, exit Route 3 and head north on Kennedy Boulevard in Union City to 49th Street, then east to Boulevard East, which follows the rim of the Palisades. Turn south and stop for a spectacular panorama of the Manhattan skyline. A small park just south of the West New York–

Weehawken boundary not only provides the stopping place and the view but also marks the site of the historic duel between Alexander Hamilton and Aaron Burr! Former transatlantic liner docks crowd the Midtown Manhattan side of the Hudson, directly across from you. The towers of Midtown Manhattan rise behind the docks, and the towers of Downtown (Lower Manhattan), four miles south, are dominant peaks on the skyline. The lower-profile area between them includes the Garment District, Greenwich Village, and a dozen other legendary neighborhoods. The fine old buildings facing the Hudson River on the bluffs north of Midtown front the famous Riverside Drive. Behind that facade, the low-profile area north of Midtown includes Harlem and the districts east and west of Central Park.

On the New Jersey side, two hundred feet below you, new pleasure-boat marinas are replacing the ruins of old railroad docks. In the heavily Hispanic and Asian neighborhoods behind you, blocks of turn-of-the-twentieth-century apartment buildings march downhill toward the Meadowlands. The landscape reflects both the historic importance of the location on the world map and the historic contrast between developments on either side of the Hudson River on the regional map.

Now it's time to continue south on Boulevard East to Route 3 and plunge through the Lincoln Tunnel into Midtown Manhattan and settle down for the night.

MIDTOWN MANHATTAN

The next day (Day Five) leave your vehicle in a parking facility and behave as most Manhattan residents—use the sidewalks and subways. Your exploration of Manhattan begins in Midtown on West 42nd Street and Seventh Avenue, at Times Square.

Forty-second Street was immortalized in a 1933 musical as the place where the "underworld meets the elite." The section from Times Square to the west is now a major center for merchandising pornography, sexual services, and drugs. The substantial homeless population is evident, and dozens use the nearby Port Authority Bus Terminal for nighttime lodging. A major office-retail-theater redevelopment is projected for this blighted space.

Midtown and Lower Manhattan

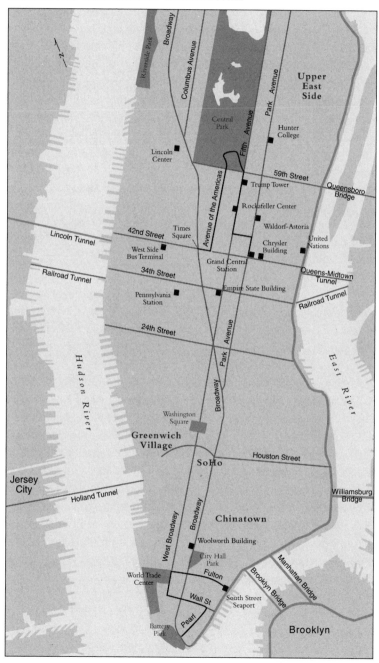

One Times Square is the location of the former headquarters of the *New York Times* and the tower from which a lighted sphere descends each year at midnight on December 31 to open the city's most traditional New Year celebration. Look north into the heart of the theater district along Broadway's "Great White Way."

Turn north on the Avenue of the Americas. In the mile to Central Park, pass impressive headquarters buildings of corporate conglomerates based notably in publishing, radio and television, motion pictures, chemicals, and finance. Small publishing offices survive from earlier, less expensive times, while branches of major foreign banks mark the new wave of multinational giants. A geographer's paradise, the main store of the Hagstrom Map Company is just east of the Avenue of the Americas on 43rd Street. Four blocks north lies the Radio City Music Hall, part of the 1930s megastructure Rockefeller Center.

Scores of individual hawkers and curbside stands contrast with the ostentatious corporate towers. Ubiquitous battalions of ready-mix concrete trucks and construction workers reflect the ongoing, colossal campaign of infrastructure maintenance, and dozens of steel plates—covering random, gaping holes in the pavement—symptomize the state of the battle.

At Central Park South, stroll a bit into the south end of Central Park, the largest park (840 acres) on Manhattan and one of the world's most famous. The park is between Fifth and Eighth avenues and 59th and 110th streets. It was proposed with great foresight in 1851. By 1856 the city had purchased some 7,500 lots which were dominated by large swampy tracts, littered with shacks, and overrun with pigs and goats. To say that this land was on the urban fringe is an understatement. There were almost no trees on the site, either. Improvements commenced in 1857 after the prize-winning design plan of (the now famous landscape architects and American heroes) Frederick Law Olmsted (b. 1822; d. 1903) and Calvert Vaux (b. 1824; d. 1895). Natural beauties of the site were retained and enhanced; 185 acres were devoted to lakes and ponds; bridle paths, walks, and roads were added; and more than half a million trees, vines, and shrubs were set out. Today Central Park is even more important as a greensward for New York

City than it was a century ago, but the park's 20 million annual visitors take a heavy toll on park management and maintenance. Monumental architecture from the entire twentieth century towers above the park and its now mature trees. Many of the world's elite, and others equally affluent, overlook the scene from their expensive neighboring apartments.

Move south on Fifth Avenue, which is still perhaps the world's most prestigious shopping street. You'll recognize the leading names of international style and retailing—names of the moment or of the century. Prices vary, but a single item could cost the same as my (or perhaps your) annual income.

The Trump Tower, south of 57th Street, is a marble-walled vertical shopping center with an 82-foot waterfall in its opulent atrium. It epitomizes the flamboyance of its namesake developer, Donald Trump, who, in turn, epitomizes those who promoted the hundred-million square feet of office-tower space added to Manhattan in the 1980s.

Between 52nd and 48th streets, the twenty-one-building complex of Rockefeller Center houses the headquarters of several corporate giants, gardens, a skating rink, entertainment, and shopping mall. A Japanese conglomerate acquired controlling interest of the complex in 1989, to the dismay of many Americans.

St. Patrick's Cathedral of the New York Catholic Archdiocese stands across Fifth Avenue at 50th Street. The age and Gothic architecture of this temple of God distinguishes it from the surrounding temples of Mammon. A mile south the 1930s Empire State Building—promoted by New York's Alfred E. Smith (b. 1873; d. 1944), first Roman Catholic candidate for the U.S. presidency—looms above the entire impressive scene. And a block south the world's pre-eminent diamond market flourishes on 47th Street.

Turn east on 48th Street, and cross Madison Avenue to Park Avenue, an elegant boulevard created above the New York Central Railroad tracks at the time of electrification. Look to the north past the prestigious Waldorf Astoria Hotel and more major corporate headquarters, into the forest of residential towers—with an understory of old mansions and townhouses—in the city's extensive "gilded ghetto" of the Upper East Side, east of Central Park below 96th Street.

Turn south on Park Avenue, and pass beside the 1960s Pan Am Tower (its namesake airline has since dissolved into a new generation of more successful American air carriers headquartered in Atlanta and Dallas) and into Grand Central Station, completed in 1913. A restored architectural landmark and once the world showplace of railway passenger service, it still serves scores of thousands of Connecticut and Westchester County (New York) commuters. Reluctantly, it was also serving in the 1990s as a nighttime lodging for some of the city's homeless.

As you look southward from Grand Central Station, Park Avenue continues as a well-maintained, older prestigious address. Also look eastward along 42nd Street to the Chrysler Building, which, along with Rockefeller Center and the Empire State Building, is one of the great architectural legacies from the 1930s. Further east is the United Nations Plaza, built soon after World War II as headquarters and meeting place of the General Assembly and the Security Council. Foreign consulates are concentrated between Rockefeller Center and the UN Plaza and north along Central Park in the Upper East Side.

LOWER MANHATTAN

Take the Lexington Avenue subway (downtown) from Grand Central Station to the Bowling Green station and Battery Park, at the lower tip of Manhattan. The park's name derives from the artillery the Americans placed here against the British during the Revolution in 1776 and again during the War of 1812. The esplanade overlooks New York Harbor and provides a view across the bay to Ellis Island, the Statue of Liberty, Governors Island, the now largely defunct Brooklyn docks, and the Verrazano Narrows Bridge between Brooklyn and Staten Island.

Cross State Street at the east edge of the park, and enter Pearl Street. Fraunces Tavern at 54 Pearl Street is a superb replica of the 1719 building where George Washington's black Caribbean steward was a proprietor and innkeeper. The tavern is the site of Washington's farewell address to his officers in 1783.

The unending process of maintenance and replacement in Lower Manhattan's venerable world financial district.

Continue northeast on Pearl Street to Wall Street, looking into the maze of curving, narrow lanes and human-made canyons. Wander westward past Federal Hall and the New York Stock Exchange. Federal Hall, built in 1849, stands on the site of the first city hall (1699), scene of President Washington's inaugural address in 1789 and the first assembly of the U.S. Congress. The ages of office towers in the neighborhood span more than a century. They symbolize the district's long-time position as North America's largest complex of banks and securities firms, and one of the largest in the world.

Here, too, the ubiquitous, makeshift steel-and-plywood patches in pavement and sidewalk and the occasional squadron of hard-hat crews and beeping backhoes remind you of the endless task of maintenance and replacement. If the country's built environment were to be fully preserved, about 2 percent of it should be in the process of rehabilitation at any time. Even in this center of wealth and power, that is clearly not the case. The gap between ideal maintenance and reality in the landscape is a measure of the overall deterioration constantly taking place. It's gradual in this area, precipitous in others less well attended.

If it's near lunchtime, you can follow the midday crowds to one of the places that serve thousands of people who already this morning have recorded or guided billions of dollars in transactions. By chance you might be led to Wolf's—a large, quaint, efficient coffeeshop located at 42 Broadway.

Nearby Trinity Church, on Broadway facing Wall Street, is New York City's oldest, chartered in 1697. On the south side of the churchyard are the graves of Robert Fulton (b. 1765; d. 1815) and Alexander Hamilton (b. 1757; d. 1804)—appropriately important names in the history of American transportation and finance.

Northwest of the burial ground, walk through the ground level of the twin 110-story towers of the World Trade Center, built by the New York–New Jersey Port Authority in the 1970s—world's largest office complex, workplace of 70,000 people. Added in the 1980s, on made land in the Hudson River immediately to the west, are the World Finance Center and Battery Park City, a publicly

Towers dating from the late nineteenth century to the modern World Trade Center provide the setting for eighteenth-century Trinity Churchyard at the head of Wall Street.

Fulton Fish Market buildings await re-use as gentrification expands around the nearby South Street Seaport.

The Battery area of Lower Manhattan from the South Street Seaport. The mast of a sailing vessel at the seaport museum and an elevated express-way are in the foreground.

financed complex of five 50-story buildings and thousands of luxury condominiums.

Return to Broadway and proceed north to Fulton Street. Two blocks north of Fulton the spired, 60-story Woolworth Building rises above the west side of Broadway. Architect Cass Gilbert (b. 1859; d. 1934) designed it to be a "cathedral of commerce." Completed in 1913, it was the world's tallest building until the Chrysler and then the Empire State buildings surpassed it during the 1930s.

Across Broadway, also north of Fulton Street, spacious grounds surround City Hall. One of the finest examples of Federal Period architecture, it has been the seat of the city's government since 1811. An adjoining complex of modern buildings is the heart of today's giant municipal operation.

Proceed eastward on Fulton Street to its eastern end, where the Fulton Fish Market still survives, though the local fishing fleet that supplied the market from 1869 onward has long since disappeared. But the ubiquitous Rouse Company (developer of Columbia, Maryland; Baltimore's Inner Harbor; and Philadelphia's Gallery Center) has restored the old fishing port area as the South Street Seaport—a collection of shops, restaurants, historical landmarks, and historic sailing ships. The ends of the piers provide close-up views of the Brooklyn waterfront and skyline, while a mix of barges, tugboats, and pleasure craft pass on the East River.

MANHATTAN'S CHANGING NEIGHBORHOODS

Return westward on Fulton Street to Broadway, and board the Lexington Avenue subway northbound (uptown). Take a local (*not* express) train, and get off at one or more stops to observe today's uses of the mass of low-profile nineteenth-century buildings and scattered new blocks between the towers of Downtown and Midtown. Three small areas provide samples of the changing variety of individuals, continually grouping and regrouping in changing geographic configurations, attaching themselves in multifarious manners and degrees to the metropolitan economy.

Get off at Canal Street, and stroll east past Centre Street into Chinatown. Heart of New York's Chinese community for more

Masonry-and-iron fire-escape facades on Broadway in the SoHo district.

than a century, with many exotic shops and restaurants, it is expanding rapidly with the influx of Hong Kong and Vietnamese immigrants.

From the next station, at Spring Street, walk through the expanding eastern edge of the SoHo ("South of Houston Street") district. Blocks of once-proud, then abandoned nineteenth-century multi-story warehouse and factory buildings, are now dominated by an expanding artist-intellectual population who could no longer find affordable space in Greenwich Village. Many small, genteel places for mid-afternoon refreshment and relaxation can be found in this district.

From the Bleecker Street subway station, walk west across Broadway and north on West Broadway to Washington Square—well worth a brief daytime stroll in the teeming heart of Greenwich Village. The eighteenth-century colonial farming village, turned

Bleecker Street in Greenwich Village.

nineteenth-century artistic-literary colony, has grown rapidly since the 1970s. The "Village" is still the center of a distinctive cultural mix of cynicism and joy, nonconformity and ordinary materialism; its population spreads over a wide range of incomes, education, and age; it is still a center for selective immigration from other parts of Megalopolis and the United States; and it is still a staging area for many who move on to other, different neighborhoods, in Megalopolis or remote corners of the country.

Continue westward on 4th Street to the 7th-8th Avenue subway station to return to your Midtown hotel. An evening dinner and stroll are in order to conclude your fifth day. The array of opportunities is incredibly rich—Midtown hotels, the theater district, the foreign consulate area, the Lincoln Center district, the Upper West Side, or perhaps the Empire State Building for a panoramic review of the geography of the inner part of the metropolis.

A crowded, decaying block of Harlem tenements in the mid-1960s . . .

. . . becomes a mostly abandoned, burned out, partly demolished block in the 1980s, with the resulting vacant land returning to woods and prairie and, in some neighborhoods, to public vegetable gardens.

UPPER MANHATTAN

Begin travel the next morning (Day Six) with a jaunt (in your vehicle) through Upper Manhattan. Follow Columbus Avenue north from the Lincoln Center area to 96th Street, through many blocks of turn-of-the-century upper-value housing, later deteriorated, and now gentrifying at the hands of a new generation of professionals.

Continue west on 96th Street to Amsterdam Avenue, then north past the Cathedral of St. John the Divine, reputedly the world's largest, and (following tradition) not yet completed after its first century of fitful construction. Adjacent is the campus of Columbia University, one of the historic Ivy League institutions, founded in 1754.

Turn northwest on 125th Street to Riverside Drive. Formerly middle- and upper-income neighborhoods, the area has seen marked racial turnover and structural deterioration in the past two decades. It is now considered to be part of Harlem, which extends to the East River and south to Central Park.

Continue north on Riverside Drive. The bluffs provide a panorama of the Hudson River and the Palisades. An immense new sewage treatment and disposal plant enriches the scene and reflects the major progress, in the first quarter-century of environmental legislation, toward cleaning up the river.

Pull onto Interstate 95, sliced at great cost through several blocks of the upper end of Manhattan. Head east across the Harlem River, and enter the Bronx.

SOUTH BRONX

One of the five boroughs of the city, the Bronx occupies the lower end of the peninsula between the Hudson River and Long Island Sound. It is a large, diverse area, home to more than 1.2 million New Yorkers. With Brooklyn, it is one of the two lowest-income counties in the twenty-three-county metropolitan area.

Urban expansion outward from Manhattan, during the latter part of the Railroad Era and early part of the Automobile Era, spread across the Bronx, as it did across northern New Jersey. But development was quite different from New Jersey. The Bronx was not a

Transient retailing and residential use in once-ornate buildings in South Bronx.

port-rail-truck gateway to the West, so its development was much less industrial and much more residential. But it was not separated from Manhattan by the Hudson River barrier, so residential density was much higher in the Bronx than in New Jersey.

The major wave of black immigration which began in Harlem in the 1910s expanded to the Bronx, mainly in the post-World War II years. An Hispanic wave, mainly from the Caribbean, has followed more recently. The new waves broke faster, on a larger scale, and with greater cultural and racial gaps than the earlier ones. In 400-acre Woodlawn, a famous cemetery in northern Bronx, graves of George M. Cohan, Victor Herbert, Oscar Hammerstein, Joseph Pulitzer, Jay Gould, and Fiorello LaGuardia testify to the varied origins of those earlier streams.

Exit south from Interstate 95 at Webster Avenue. Continue east three blocks on 174th Street, and turn south through the Bathgate

Upper Manhattan – South Bronx

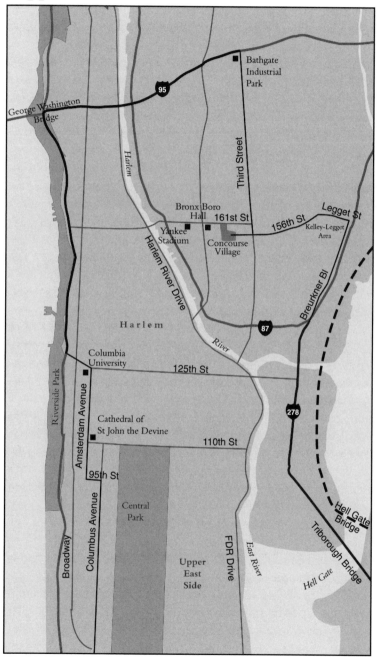

Industrial Park. Located near the freeway, this major investment by public agencies is an effort to attract employers and create jobs for the under-employed labor force of the South Bronx.

Bathgate is but one part of a many-sided approach to improve local economic opportunity, education, health, housing, and household management. The needs are great. In 1988, the Bronx Economic Development Council reported 200,000 adults who could not read and write at a fourth-grade level; severe drug dealing and crime problems; thousands of newly-arriving families with social, psychological, and economic problems; and housing abandoned by private owners who could not maintain the buildings under rent control, then taken over and poorly maintained by the city. Health care and social services were swamped by the human tidal wave. The Council noted the lack of trees and the trash and junk that give the borough a "dumped-on" image. Yet, adjoining Bathgate on the northeast, another of New York City's "Little Italy's" holds its own amidst the surrounding blight and offers an exciting enclave of restaurants and shops.

Leave the east side of the Bathgate development and head south on Third Avenue to 156th Street. The landscape testifies to the depth of recent problems. But it also speaks of past communities, and of thousands of current residents building and maintaining local communities and connecting to the wider metropolitan economy.

At 156th Street go west three blocks to Concourse Village Houses to see one of the many publicly financed approaches to replacing the deteriorated housing stock. The monumental Bronx City Hall—with security problems even inside the building—is one block west, on Grand Concourse. Author Tom Wolfe's *Bonfire of the Vanities* describes this area in a gripping and colorful style.

Return eastward on 156th Street, across Third Avenue to Leggett Avenue. The next intersection, where Leggett crosses Kelly Street, is the heart of a grass-roots neighborhood rehabilitation project, which succeeded in a more comprehensive community approach to housing deterioration.

Continue southeast on Leggett Avenue to Bruckner Boulevard, then south to the entrance to Interstate 278, and head south across the Triborough Bridge (begun in 1929 and completed in 1936).

Toward the east and northeast are several extensive land uses essential to the operation of Manhattan but exiled to the margins. Major sewage treatment and electric power plants face the East River. A prison occupies Rikers Island. North of that, on the low peninsula between the East and Bronx rivers, is the extensive Hunts Point produce market. Shifted from the much more expensive land on the Lower East Side of Manhattan in the 1950s, the market handles three-fourths of all the produce imported into metropolitan New York. The shipments originate in forty-nine states and thirty foreign countries.

The Hell Gate railroad bridge parallels the Triborough freeway bridge over the East River. Unpainted since 1940, the rusting bridge exemplifies some basic problems in maintenance of the area's infrastructure—fragmented management responsibility, limited funds without rational priorities, and changing geography of industry and transportation.

QUEENS

Descend from the Triborough Bridge into Long Island and Queens. Another big borough, with a population of nearly 1.9 million people, Queens surrounds Brooklyn on the north and east. Hence, its land is generally farther from the early main crossings to Lower Manhattan, and its large-scale settlement came slightly later than Brooklyn's. On Long Island's glaciated surface, the physical diversity of Queens is less than that of the Bronx. But the cultural diversity is greater. In addition to burgeoning numbers of blacks, the extremely heterogeneous non-black population exceeds 1.1 million.

You'll notice abundant opportunities for refreshments and rest breaks, and a lunch stop, in the expedition through the ethnic neighborhoods of Queens.

From the Triborough Bridge continue east, then south, on Freeway 278 to the Queens Boulevard/Route 25 exit. Go west on Queens Boulevard to 39th Street, and proceed north across Skillman Avenue into Sunnyside Gardens. This planned community

Bronx, via Queens, to Larchmont, New York

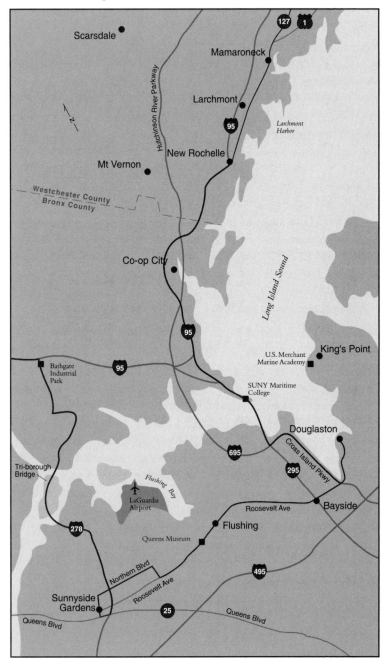

was built between 1924 and 1928 on surplus land of the major rail yards on the Hell Gate-Midtown connector. It was an American adaptation of the British garden suburbs laid out by Sir Ebenezer Howard (b. 1850; d. 1928). Through the limited-profit corporation that developed it, Sunnyside was the first project of the Regional Plan Association—a missionary group of architects, engineers, and social scientists committed to improving urban environments. Legendary planners Lewis Mumford (b. 1895; d. 1990) and Clarence Stein (b. 1882; d. 1975) were involved. Greenbelt and Columbia, Maryland, followed the same tradition, and the Regional Plan Association still plays an important civic role in metropolitan New York. Conceived as a worker's paradise, Sunnyside Gardens has been sustained by older middle-class professionals and younger "yuppies" who prize its ambience and proximity to Manhattan.

Leave Sunnyside Gardens on Skillman Avenue, east to 54th Street, then north to Northern Boulevard. Proceed east on Northern Boulevard through the Jackson Heights district to Junction Boulevard.

In addition to African-American enclaves, the business establishments indicate the ethnic mix of the neighborhoods along this two-mile stretch. A recent count included Afghani, Ethiopian, Thai, Vietnamese, Burmese, Indian, Pakistani, Chinese, Japanese, Korean, Greek, Turkish, Venezuelan, Mexican, Colombian, Brazilian, Haitian, Jamaican, Cuban, Italian, Spanish, and Portuguese. On the corner of Northern and Junction boulevards, the public school is named for benefactor Louis Armstrong (b. 1900; d. 1971), the famous musician who was a native of the neighborhood and never lost his affection for it.

LaGuardia Airport is a mile to the north, on made land along Flushing Bay. Kennedy Airport, eleven miles to the south, is on made land beside Jamaica Bay. Both are in Queens, on the north shore and south shore, respectively.

Turn south on Junction Boulevard, then east on Roosevelt Avenue. The elevated railroad overhead—like the high-density part of the metropolis it still serves—is part of the legacy from the pre-auto era. The street beneath the "El" is a thriving, colorful beehive

Ethnic and cultural landscape of northern Queens.

of activity in a neighborhood mix of Jews, Palestinians, Koreans, Chinese, Colombians, Argentinians, and Indians.

Continue east on Roosevelt Avenue into Flushing Meadows-Corona Park, and turn right to the Queens Museum, located in a building that housed the New York City exhibit in the 1964–1965 World's Fair. Inside is a 9,300-square-foot, three-dimensional model of the city on a scale of 1:1,200, or about five feet on the model to one mile in the real city.

Every building and street is there. You can circle the platform above the model and use a telescope to find the streets and buildings you have seen on this trip. The model's thousands of buildings cost $672,000 to create in 1964. Major new structures were added in 1990 and 1991. At that time, single large buildings cost $6,000 to $10,000. No older miniature buildings were rehabilitated; they just continue to age.

In short, like the real city, the model is very large and endlessly fascinating; it's an expensive accumulation of buildings; costs are going up; and maintenance is deferred. Truly a model of the real city, as well as a stimulating place to review its geography!

Return to Roosevelt Avenue, and continue east past Shea Stadium (home of the New York Mets professional baseball team) to the intersection with Main Street, in downtown Flushing—now one of the city's most flourishing retail nodes as a result of a very large East Asian influx, and despite large-scale white flight eastward to neighboring Bayside.

Continue eastward on Roosevelt, then on Northern Boulevard beyond Little Neck Bay to the entrance to Douglaston. Turn north into a neighborhood of large single-family homes, mostly of twentieth-century vintage. A farm until the late nineteenth century, Douglaston is built on hummocky glacial moraine and bluff land jutting into Long Island Sound. Its residents are cautious about the integrity of the physical environment and the single-family residential zoning. Resistance to central sewers has kept this area dependent on individual septic tanks. But it also has kept it safe from high-rise or commercial development.

Be sure to get out to Shore Road for a view of Long Island Sound. The deeply indented shoreline, wooded bluffs, and fine homes are characteristic of much of the urbanized north shore of Long Island.

Return to Northern Boulevard, and go west a short distance to the Cross-Island Parkway, then north along Little Neck Bay to Interstate 295 and the Throgs Neck Bridge. This span at the end of Long Island Sound was yet another piece in the difficult development of the metropolitan circumferential freeway system.

ANOTHER FACE OF THE BRONX

As you cross the Throgs Neck Bridge, the State University of New York Maritime College (founded in 1874) is below you, and the U.S. Merchant Marine Academy (founded in 1938) is to the east across the bay. Both are reminders of the historic importance of maritime shipping to Megalopolis, and the precarious position of American flag shipping in the face of world competition.

Beyond the bridge, keep right on the Interstate 695 cutoff to Interstate 95. Continue north on I-95 through twentieth-century residential districts in the easternmost section of the Bronx, past the monstrous cluster of towers called Co-Op City, built on flat, made land—once a bankrupt amusement park—at the mouth of the Bronx River.

Opened in 1968, Co-Op City was developed under a state financing assistance program. Its 15,000 apartment units are home to 50,000 people, with black, Asian, and Hispanic minorities now heavily represented. In the former Soviet Union, rental agents there facilitate Co-Op City rentals and immigration for Jewish families. Picture a geography of the global social networks focused on this mass of towers!

WESTCHESTER COUNTY, NEW YORK

Resume the northward traverse on Interstate 95 into Westchester County. Westchester and adjoining Fairfield County, Connecticut, occupy the wide, upper part of the peninsula between Long Island Sound and the Hudson River.

Three trunk rail lines have long provided access to Manhattan— one line along the Hudson shore, another connecting the bayheads on Long Island Sound, and a third on the Westchester County upland. Early in the twentieth century, compact suburban communities grew near the railroad stations. Population in these suburbs exceeded 900,000 on the eve of World War II. But the intervening land was mostly rolling, wooded, and pastoral countryside and expansive country estates of wealthy elite. Large parts of the Westchester County watersheds had long been protected and dammed to supply water for New York City.

After World War II, a vast sprawl of affluent households in subdivisions and on small acreages burst over the scenic inland hills. By 1990, the two-county population was 1.7 million.

Many commuters began to take advantage of modern communications and computer technology and shift their business to the vicinity of their homes. New start-ups and established firms began to move to Westchester County. Headquarters/office parks of numerous major corporations are only the most obvious manifestation.

In western New Rochelle, exit Interstate 95 onto eastbound U.S. Route 1—in this area still named for its colonial predecessor, the Boston Post Road. The highway skirts the bays that shelter marinas and yacht clubs and open to Long Island Sound. Long promontories between the bays accommodate spacious neighborhoods of homes. An excellent example is Larchmont Harbor and the residential areas on either side of it.

Follow Route 1 through New Rochelle, Larchmont, and into Mamaroneck. Angle northeast on Route 127 into the town of Harrison. The Harrison railroad station is surrounded by a rectangular street grid, small blocks, and Railroad Era houses. Then, a short distance north, you enter the long-block, curvilinear world of affluent post-World War II suburbia, which dominates the upland landscapes of the peninsula between Long Island Sound and the Hudson River.

Follow Route 127 on its winding, hilly route northward past fine homes and country clubs to White Plains. Turn west on Westchester Avenue, and circle the opulent downtown multi-block retail-office-hotel redevelopment, which offers one possible choice for a mid-afternoon rest stop.

Return to Westchester Avenue, cross Interstate 287, and continue east on Anderson Hill Road through the village of Purchase. The General Foods corporate headquarters park is across the interstate to the south. Pass the campuses of Manhattanville College (founded in 1841) and the State University of New York at Purchase (founded in 1967), and three miles east of Interstate 287 turn south into the Pepsico headquarters and Public Sculpture Garden.

From Pepsico Gardens head south on Lincoln Avenue through spacious residential areas, across the Hutchinson River Parkway, and notice the Texaco corporate headquarters park a short distance southwest. The array of office parks in this area is only a small sample of the vast development along Interstate 287 on the northern and western metropolitan fringe.

Turn north on the attractive Hutchinson River Parkway. At the Connecticut state line, the "Hutch" becomes the Merritt Parkway. Continue eastward on the "Merritt" through the back country of

Larchmont, New York to Stamford, Connecticut

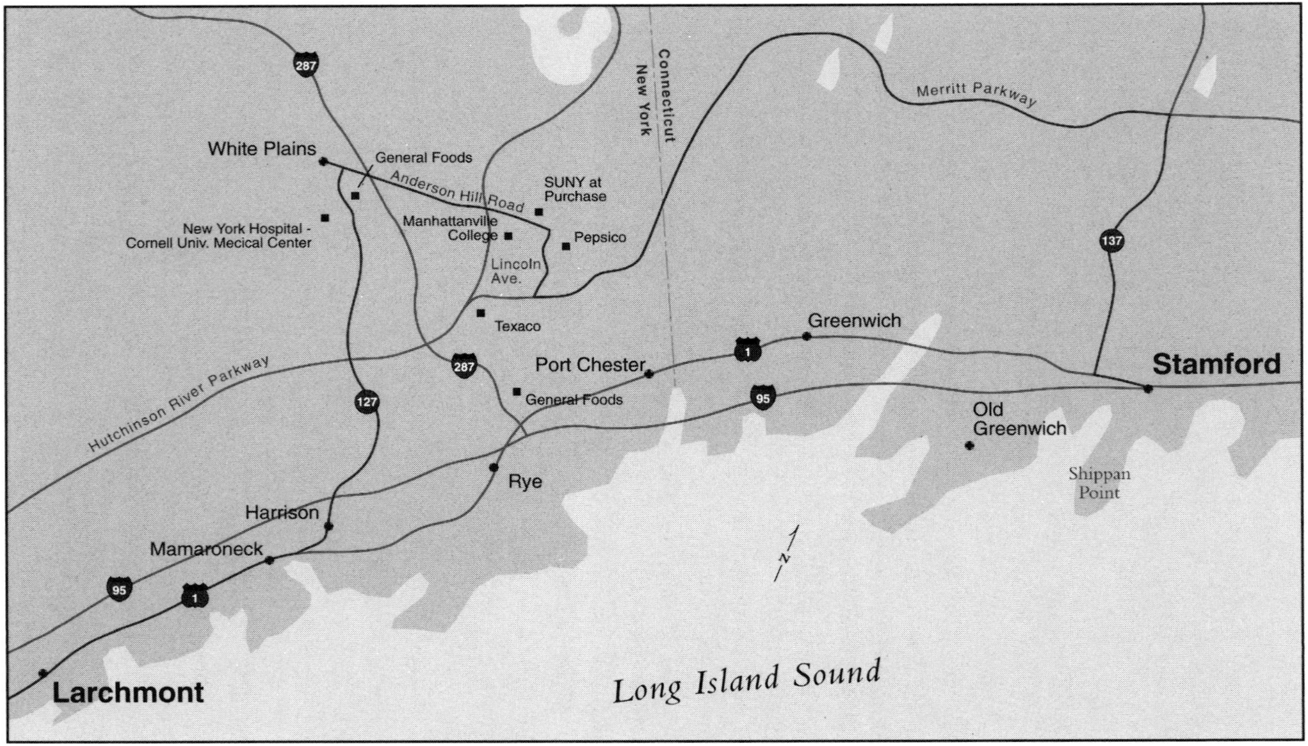

Post-World War II high-value homes fill the rocky woodlands of Westchester and Fairfield counties in New York and Connecticut.

Turn-of-the-twentieth-century mansions dominate many peninsulas on the Connecticut suburban coast of Long Island Sound, such as this home in Greenwich, Connecticut.

The new (post-1960) landscape of millionaire suburbia and transplanted Manhattan offices overwhelms the colonial harbor of Stamford, Connecticut.

Greenwich eight miles to Route 137, then south into central Stamford to spend the night (end of Day Six).

Stamford (settled in 1641) has been a commercial sailing port and fishing center, an industrial center, a commercial center for the suburbanizing southwest corner of Connecticut, and, since the 1970s, the center of major downtown high-rise office development for Manhattan expansion and dispersal. Around the inland edge of the glittering downtown, a substantial minority population occupies workers' cottages remaining from the industrial period. Meanwhile, Shippan Point—a long promontory in Long Island Sound once reserved for a park—was developed for private estates after 1890, and marinas for pleasure craft occupy the old port. Inquire about possible places for dinner in the partly rehabilitated, old industrial district on the harbor.

△ Day Seven

ACROSS CONNECTICUT AND RHODE ISLAND

Guideposts along the Route

GRANDEUR AND GRITTINESS

Leave Stamford on Interstate 95 (north) and continue to parallel the shore of Long Island Sound, past Norwalk and Bridgeport. The Connecticut shore urbanized very early. By 1840 three-fourths of the population lived in towns of 2,500 or more, and this figure grew by almost 100 percent by 1880. The nearby New York metropolis catalyzed industrial development in the nineteenth and early twentieth centuries; and it infuses office, services, and metropolitan residents today.

Norwalk replicated Stamford's pattern of dense settlement near the Sound, but with a more diversified industrial base. The city has been less attractive than Stamford for new office and commercial development, and recent population growth has been slower.

Beyond Norwalk, Interstate 95 cuts through predominantly affluent suburban Westport and enters the much older, compact, somewhat distressed industrial city of Bridgeport. One recent issue of the local newspaper reported that TV personality Phil Donahue had purchased a two-acre residential property in Westport for $4.8 million, while the same issue reported that Bridgeport officials—struggling with problems of maintaining an old infrastructure and providing services on a much lower per-capita tax base—had filed to ask the courts to declare the city in bankruptcy.

Stamford, Connecticut, to Boston, Massachusetts

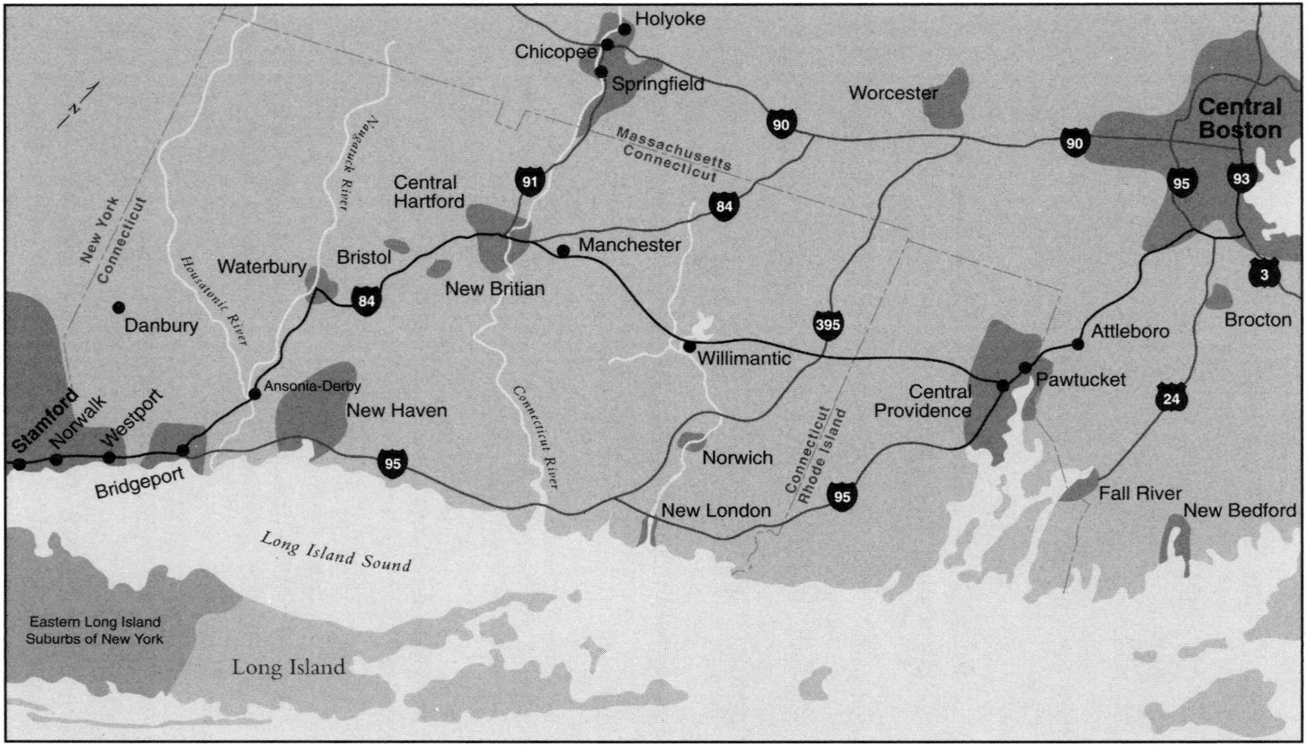

Historically the dominant industrial city on Long Island Sound, Bridgeport had 20,000 manufacturing employees in 1900 working in large firms like Bridgeport Brass, Remington Arms, Union Metallic Cartridge, a large sewing machine company, and many small firms making producer durables and hardware. Following a boom during World War I, many firms focused on products for the automotive and electrical industries. But recovery from the Great Depression again depended mainly on military demand. A new industry, Sikorsky Aircraft, was a notable beneficiary. Bridgeport's gritty, industrial character has hampered revitalization in recent decades. Its population declined slightly in the 1980s, to 141,000.

THE HOUSATONIC AND NAUGATUCK RIVER VALLEYS

Exit Interstate 95 in central Bridgeport, and head north on State Route 8. Cross the Housatonic River and follow the tributary Naugatuck River north through the old industrial water-power cities of Derby and Ansonia. Large brass and copper mills, financed and founded in part by New York City investors, were developed here by 1850. Ansonia was named for Anson Phelps, a Naugatuck Valley resident who became a successful metals trader in New York and founded some of the early mills. Phelps Dodge, a major international copper company, had its roots in his efforts. These cities have always been small, specialized towns, focused on their factories. The factories and nearby workers' neighborhoods and churches dominate the hilly landscape. Flooding in the narrow valley has always been a problem, and today's diked channel dramatically constrains land use.

At the Housatonic River the route at last leaves the New York Consolidated Metropolitan Statistical Area. As the route continues through Connecticut, it crosses the northern part of New Haven-Waterbury (804,000 population), then three-county metropolitan Hartford (852,000 population), and finally the only county on the entire Megalopolis traverse that is not in a census metropolitan area—Windham County (104,000 population) in Connecticut's northeast corner. This is also the transition zone of commercial dominance between New York City and Boston.

This water-powered textile mill has been converted to a gracious office complex near Danbury, Connecticut.

Leave the freeway for a stop at the Mattatuck Museum in downtown Waterbury for excellent exhibits on the brass industry and Waterbury's history, as case studies of New England manufacturing and urban development. Early metal working and clock making in Waterbury led to a boom in brass manufacturing, powered by the Naugatuck River. Fifty thousand people were employed in brass-related industries at the zenith of the "Brass Valley." Irish, Italians, and French Canadians dominated the highly ethnic population. They made Waterbury the center of triple-deck residential architecture in Connecticut.

Damaged by severe floods in 1955 and a hundred-million-dollar tornado loss in 1989, the city is suffering longer-term ills of cities that are heavily dependent on industries in decline. The decline is counteracted by growth in new industries, reflecting the continued outward spread of New York City's economy. With a population of 109,000 in 1990, the city had grown by 6 percent in a decade.

WATERBURY TO HARTFORD

Follow Interstate 84 east from Waterbury. About ten miles toward Hartford the freeway skirts Southington Mountain and drops to the Connecticut Lowland. As it bypasses the centers of Bristol and New Britain, the route becomes the axis of a major development corridor—new houses, apartments, the million-square-foot West Farms Mall, corporate offices, and highway-oriented small business. Exit at downtown Hartford for lunch and a stroll. Numerous places for lunch can be found in the modern Civic Center and the refurbished Brown-Thompson Building.

Settled in 1635 by immigrants from the Massachusetts Bay Colony, Hartford's early growth was based on its location in the heart of the Connecticut lowland and the head of eighteenth-century ocean sailing on the Connecticut River. The 1879 state capitol building, crowning a hill at the western edge of downtown, and the 1796 old statehouse, nearer the river, reflect the city's

Hartford's impressive downtown skyline on the Connecticut River.

historic importance in Connecticut. The insurance industry dates from the late 1700s, and generations of entrepreneurs took advantage of the early start and Megalopolitan location to make Hartford a major insurance center.

Large-scale central area redevelopment has enhanced an already impressive mass of office towers and public buildings. City Place, the state's tallest office tower, symbolized the importance of corporate offices to the Hartford economy. The new Prudential Plaza and the Phoenix Insurance Building, which replaced old factories and residential slums, speak to the importance of both insurance and post-World War II redevelopment. New high-rise apartments on the flats west of the capitol occupy land that was the center of the U.S. typewriter industry until the 1950s. Retail redevelopment has not been successful, and downtown office vacancy has been exceptionally high. Many insurance company back-offices have relocated to the suburbs, where some stand in the former tobacco fields for which Connecticut lowland agriculture was famous.

The city of Hartford has only about one-sixth of the metropolitan population, but it includes most of the black and recently arrived Hispanic minorities, who have occupied the central, northern, and western neighborhoods. A residual Italian community lives in the southwest. Virtually all growth has been displaced to the suburbs. High turnover and minority concentration give the city some of the highest poverty and infant mortality rates in the nation.

Continue east across the Connecticut River on U.S. Route 6 toward Willimantic, Connecticut, and Providence, Rhode Island. At the east edge of downtown Hartford, the stone-arch Bulkeley Bridge marks the historic head of navigation. To the south, the distinctive dome of the Colt firearms company is visible. East Hartford is the home of United Technologies and its subsidiary, Pratt and Whitney, the world's largest marker of aircraft engines. Both are reminders of the long-time importance of defense-related industries in and near the Connecticut River valley. Vulnerable to defense-spending cuts and shifts, East Hartford lost ten thousand jobs in the 1980s.

Dikes protect the extensive floodplains in the East Hartford area. Disastrous floods truck in 1936 and 1938. Most of the post-World War II development occupies former prime agricultural land.

Keep to Route 6 through Manchester, and note the former Cheney Silk Mill—an example of adaptive reuse of old industrial buildings. Silk was first made here in 1838, and the enterprise grew to five thousand jobs, thirty-six acres of floor space, and a two-mile independent railroad. The surrounding workers' village was an exception to the industrial feudalism common to nineteenth-century New England—the workers were well paid and well housed, and child labor was never allowed.

Suburban expansion in the area has helped Manchester's business district to survive in spite of the development of the state's largest shopping mall on the bypass.

HISTORY AND HIGHWAY IMPACT IN THE "QUIET CORNER"

Leave the Connecticut lowland at Manchester and continue east on Route 6 to Willimantic, a small industrial city incorporated within the town of Windham. Known as the "Thread City," it reached its heyday in the early twentieth century. On an upper tributary of the Thames River, its water-power-based industries made textiles and textile machinery.

Like New England industrial cities of all sizes, Willimantic struggles with maintenance and adaptation of its buildings, infrastructure, and community structure in the face of population turnover and slow growth. Hispanic immigration has swollen, following early labor recruitment by the American Thread Company. Meanwhile, like larger cities, Willimantic used federal grants and local funds in the 1970s in efforts to change the look of the place. Old industrial plants and related neighborhoods were razed, but much of the land stubbornly remained vacant. Much of the housing stock was upgraded, and the Hill Section is now an attractive residential neighborhood. When the big American Thread Company complex closed several years ago, its adaptive reuse became the community's goal.

Willimantic's Windham Textile Museum provides an opportunity to combine a mid-afternoon rest stop with an informative review of the industrial basis for this part of the country from a knowledgeable director.

Continue east on Route 6 across Windham County and the "quiet corner" of Megalopolis. At the time of the first national census in 1790, this area had the lowest population density in the Boston-Washington corridor. And it still does—with an exurban landscape of mostly blue-collar population living in the interstices of metropolitan Hartford, Worcester, Boston, Providence, and the Trident submarine/General Dynamics plant at Groton–New London.

While high-wage jobs in Groton had long attracted the local labor force, the area became attractive for immigrant long-distance commuters during New England's economic boom of the 1980s because of relatively low land and construction costs. Only a little of the new housing is visible from the road; most nestles in the woods. The resulting increase in service jobs for the local population is most evident along north-south Interstate 395. Large distribution warehouses and bulk packaging plants have recently appeared in that corridor, where they are highly accessible to all of New England. Visible problems of traffic and waste management reflect the exurban development.

METROPOLITAN RHODE ISLAND

Enter metropolitan Providence at the state boundary, and follow Route 6 to the center of the city. The metropolitan area embraces almost the entire state. The Colony of Rhode Island and Providence Plantations was chartered in 1663. Urban growth concentrated in Newport until after the Revolutionary War, then Providence emerged as the commercial leader. However, it has always existed in the shadow of Boston and New York City, in a powerful transportation corridor—first ship, then rail, now freeway. Today, commuting links the north side of metropolitan Providence with both downtown Boston and the high-tech jobs on the Boston circumferential freeways (Interstate 495 and Route 128).

Population swelled during the nineteenth century with the growth of the textile industry, on the Blackstone River in the northern suburbs, and on the little Woonasquatucket River in the city. The industry peaked in the early twentieth century, then declined precipitously, with southern (and later, foreign) competition. Relict mills dot the city today, especially in the Olney district, where today's Route 6 nears the Woonasquatucket west of downtown. The machine-tool and jewelry industries also became important in the nineteenth century.

Exit at the Civic Center, and pick your way through the redeveloped and rehabilitated downtown. Downtown retailing serves partly the inner-city, low-income market and partly the daytime white-collar office workers. City leaders hope the attraction of lower labor costs and rents will add to the existing banking and business service base.

At the east edge of downtown, cross the narrow Providence River to the vicinity of College and Water streets on the river's east bank. A major public-works project has created a promenade along the uncovered channel of the 1828 Providence and Worcester Canal. The promenade affords a fine panorama of the mix of old and new facades and towers of downtown, and of early buildings on the east bank.

Circle around the campus of historic Brown University (founded in 1764) and the adjoining old elite, ridgetop residential district.

Return to Interstate 95 north of downtown, and head north through Pawtucket toward Boston. In the compact, industrial satellite city of Pawtucket, the freeway crosses the Blackstone River, where it runs through a corridor of monumental nineteenth-century textile mills and just below the historic Slater mill—one of the earliest in the U.S., with an excellent restoration and museum.

Beyond Pawtucket the route crosses the Rhode Island–Massachusetts boundary and enters a low-density, auto-era landscape whose circulation patterns focus on metropolitan Boston. (Note: We will spend the night in central Boston. Please refer to the next chapter for directions.)

Boston Region

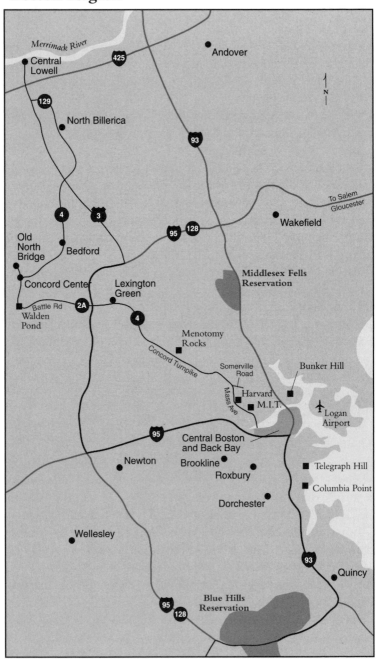

Merrimack River

425

Central Lowell

Andover

129

93

North Billerica

To Salem Gloucester

4

3

95 128

Wakefield

Old North Bridge

Bedford

Middlesex Fells Reservation

Concord Center

Lexington Green

2A

Battle Rd
Walden Pond

4

Menotomy Rocks

Concord Turnpike

Somerville Road

Bunker Hill

Mass Ave

Harvard
M.I.T.

Logan Airport

95

Central Boston and Back Bay

Newton

Brookline

Roxbury

Telegraph Hill

Columbia Point

Dorchester

Wellesley

93

Quincy

95 128

Blue Hills Reservation

N

$\triangle$ *Day Eight*

METROPOLITAN BOSTON

Prologue

The 2,400-square-mile Boston Consolidated Metropolitan Statistical Area of the U.S. Census includes five counties and covers most of the eastern one-third of Massachusetts. The metropolis overlaps a part of the Narraganset lowland, but most of it spreads across the stony, rough, glaciated hill country, rocky headlands, and intermittent beaches of the coastal northern Appalachians. Most of the inland area drains to the Merrimac River or the Charles River. The random collection of glacial features makes a richly varied physical setting.

At the heart of the area, roughly inside today's circumferential freeway Route 128, a lowland basin guides the Charles and two even smaller rivers—the Neponset and the Mystic—into the wide estuaries of Boston Harbor. The Neponset River widens into Dorchester Bay; the Charles and Mystic converge in the Boston Inner Harbor. Many isolated drumlins—oval-shaped, rock-cored, smoothed piles of glacial material—rise steeply out of the lowland. Bunker Hill, in Charlestown, and Beacon Hill, downtown, are the most famous of these drumlins. Intervening surfaces are comprised of hummocky glacial moraine, studded with ponds, lakes, and wetlands. Or they are flat, sandy outwash plains, spread by torrents of glacial meltwater with names like Jamaica Plain. The crystalline Blue Hills form an exceptionally prominent ridge near the southern side of the basin. Hills drowned by rising sea level since the Ice Age form many islands and promontories in the harbor.

Today's metropolitan area was the heart of the Massachusetts Bay Colony in the seventeenth century. Wagon roads fanned out to the agricultural towns and water-power mill sites along the valleys. It was the heart of the system of industrial satellites focused on Boston in the Railroad Era. Mid-nineteenth-century rail lines radiated from the "Hub City" to the industrializing water-power cities. In the Steel Rail epoch they became trunk lines to Canada, the West, and southern Megalopolis, and the Merrimac Valley cities of Lawrence and Lowell became metropolitan centers in their own right. Then the region filled and coalesced to form today's consolidated metropolitan area.

The five-county region boomed in the Railroad Era, as the market for its textiles and metal goods expanded across the country. Opening of the Middle West also dried up the New England farm labor surplus, while immigrant labor streamed in from southern and central Europe and from French Canada.

After its prolonged nineteenth-century industrial surge, the region grew slowly in the first half of the twentieth century, because of the emigration and decline of its textile industry. With limited immigration and turnover, many immigrant neighborhoods stabilized to give an air of permanence to the cultural geography that turned out to be unwarranted in the long run. The slow growth also helped the legacy of historic landscapes to remain largely untouched by the 1920s building wave.

A resurgence came during and after World War II, as a result of the boom in the locally incubated electronics industry. From 1950 to 1990, the population grew from 3.1 to 3.6 million. In the same four decades, Suffolk County—mainly the Railroad Era central city of Boston—declined from 900,000 to 660,000. Meanwhile, as in all of Megalopolis and the nation, minority population grew and concentrated mainly in the central city; black population in Suffolk County, for example, increased from 5 percent in 1950 to 43 percent in 1990.

The area's distinctive growth history—nineteenth-century industrial boom, long period of slow growth, and recent resurgence—is registered in the housing landscapes. A strikingly large share of the metropolitan area is compact, pre-1920 city. Another very large share is post-1950 suburbia. There are accompanying massive needs both to organize urban sprawl and to maintain, adapt, and rehabilitate the inner city.

Guideposts along the Route

FOXBORO TO BOSTON HARBOR

Heading toward Boston, Interstate 95 intersects the outer circumferential freeway at Foxboro, Massachusetts. Completion of the two freeways created a new central location here, in sparsely settled country, yet midway between Boston and Providence and easily accessible to a population of 8 million in eastern Massachusetts and Rhode Island. The location attracted the professional football stadium of the New England Patriots.

At the junction with the inner circumferential freeway follow Interstate 93 around the southern and eastern side of the Blue Hills. Like several extensive rocky highlands in the metropolitan area, the Blue Hills have been preserved from development.

East of the Blue Hills the route turns north through Quincy—an early satellite industrial port, birthplace of John Adams (b. 1735, d. 1836), John Quincy Adams (b. 1768; d. 1848), and John Hancock (b. 1737; d. 1793), and site of major shipbuilding for two world wars.

Beyond the Neponset River the compact, streetcar neighborhoods spread northward toward central Boston. Savin Hill and Telegraph Hill drumlins rise beside the harbor to the east, and between them Columbia Point juts into the harbor. The point is the site of one of the country's notorious urban high-rise ghettos as well as the newer Kennedy Library and state archives. The expanding, predominantly black neighborhoods of Roxbury and Dorchester lie to the west and northwest, with the office towers of the Back Bay area in the background to the northwest and the skyline of downtown Boston ahead to the north.

The quiet historic waterfront and massive new office towers of Boston's financial district.

As you approach downtown, the general cargo and container wharves of the modern harbor lie to the east. Smallest of the East Coast overseas ports, Boston's container traffic is equal to about one-fourth of Baltimore's, one-fifth of Norfolk's, and 7 percent of New York's. Across the Inner Harbor in East Boston, like other Megalopolitan airports, Logan International spreads over a vast area of made land. The impressive South Station still serves as the rail terminus for the Northeast Corridor and for commuters from the southern and western suburbs.

After a panorama of downtown and the historic Inner Harbor waterfront, exit the expressway at Causeway Street, and head for your Boston overnight accommodations. Central Boston offers unmatched opportunities for looking and reflecting while strolling.

The plaques and landscapes along the well-marked, three-mile "Freedom Trail" from the Boston Common and the State House to

Two centuries of residential maintenance and preservation on Beacon Hill.

Faneuil Hall provide a good focus. If possible, add a detour west of the State House into the narrow side streets of Beacon Hill—at least to Mount Vernon Street, Louisburg Square, and Pinckney Street—for the flavor of its eighteenth- and early nineteenth-century housing and the mix of residents who enjoy living in the area and preserving it today. Another detour into the massive post-1960 Government Center provides an impressive view of the power and limitations of large public expenditures to rebuild a venerable city.

As you follow the Freedom Trail, pause along the State Street edge of the financial district to contemplate the two-century age spread of the building facades and to reflect on the immense role of locally controlled, historically accumulated wealth in the continuity of a city or a region. Compare the Rouse-developed Quincy Market festival-restoration area with the company's inner-city projects in Manhattan, Philadelphia, and Baltimore.

Consider a cruise on the harbor for a view of the impressive, mostly post-World War II skyline and at least a glimpse of the USS *Constitution* moored across the harbor, at the foot of Bunker Hill in Charlestown.

CAMBRIDGE AND SOMERVILLE

The final day of your excursion begins with a traverse through two of the world's citadels of higher education. Follow the Embankment and Memorial Drive, along the south side of the Charles River, westward to the Harvard Bridge. The buildings of the Massachusetts Institute of Technology (MIT), founded in 1861, line the opposite side of the basin. To the south, the route skirts the famous Back Bay district, whose tight rows of gentrified nineteenth-century townhouses follow Beacon Street and Commonwealth Avenue across a large area of made land on the former marshes west of Beacon Hill.

Cross the Harvard Bridge, and follow Massachusetts Avenue through the MIT campus. Its monumental structures along the basin, backed by blocks of industrial-type laboratory buildings, are the center of a remarkable network of alumni and faculty linked to all the world's major centers of engineering sciences.

Proceed to Harvard Square—intellectual-cultural heart of the Boston region, with a remarkable concentration of bookstores, theaters, concert halls, galleries, newsstands, and shops. (A colleague from another university assures me that the concentration is "unmatched in eastern Massachusetts"!)

Here, in a Puritan village named Newtowne in 1636, the Massachusetts Bay Colony established a college to train ministers for the newly settled towns of the colony and decreed that Newtowne be renamed Cambridge. Since that time, Harvard's network of graduates and faculty has reached into the most important centers of influence throughout Megalopolis, and through them to other centers worldwide.

Adjoining the square, Harvard Yard is the ancient core of Harvard University, a giant city block now devoted to freshman undergraduate dormitories, Widener and Houghton libraries, Memorial Church, and the central administration building. Nearly every architectural style found in the eastern United States can be seen in the Yard. The oldest structure, Massachusetts Hall, was built in 1720 with public funds, and several nearby halls were built with money raised by colonial lotteries.

Equally eclectic but more recent buildings, mainly for science, law, and special-purpose institutes, march a half mile to the north of the Yard. A campus extension south to the Charles River includes dormitories and the John F. Kennedy School of Government, and land across the Charles accommodates the Graduate School of Business Administration and the stadium.

From Harvard Square, continue northwestward on Massachusetts Avenue a scant mile to its convergence with Somerville Avenue. A short detour is in order here, north on Davenport Street, and into side streets such as Burnside or Banks, between the arterials of Elm and Summer streets. You have crossed the boundary from Cambridge into Somerville—historically the more blue-collar of the two—and you have entered a landscape of two- and three-story frame houses distinctive to industrial New England and common over many square miles of the metropolitan area. The white pine forests of Maine were cut over to supply the material, and the buildings sheltered many a late nineteenth- and early twentieth-century immigrant family.

THE NORTHWEST SUBURBS AND HALLOWED GROUND

Return to Massachusetts Avenue, continue northwest to Alewife Brook Parkway, and jog west onto the Concord Turnpike (Route 3). Note the multi-level park-and-ride garage of the new Alewife subway terminal at the Concord Turnpike. The station marks the end of a recent extension of the century-old Downtown-Harvard Square Line, with turnpike access to the western suburbs. The extended line and the parking facility are a part of the Massachusetts Bay Transportation Authority's large, expensive effort to link the transportation systems in the old, high-density city and the newer Automobile Era suburbs.

Exit the Concord Turnpike at Route 60, and follow Old Spring Street and Eastern Avenue up the hill to the large, open park beside

Eastern Avenue, a long three blocks northwest and above the main tract of Menotomy Rocks City Park.

As far as Arlington, the route has been crossing the Boston Basin—underlain by relatively soft sedimentary rocks, and locus of much of the densely built-up city. Here in Arlington Heights, the land rises abruptly, from the basin to the surrounding Appalachian uplands. The hard, crystalline rocks that underlie the uplands are exposed at a few places in the park.

Walk to the open hilltop for one of the finest panoramas of Boston. The skyline of downtown and Back Bay stands out clearly. With one exception, all of the office towers have been built since 1950. Perhaps because of New England's prolonged economic slump at the time, Boston's skyline was not swept upward in the skyscraper-building boom of the 1920s.

Return to the Concord Turnpike, and continue westward from Arlington into Lexington. Near the town line, note Wilson Farm—one of the few remaining market-garden farms in the state. Numerous small pockets of deep soil, like the one at Wilson Farm, interrupt the rocky uplands and provided the land resource for pioneer farming communities.

Leave the Concord Turnpike at Route 4 and head northwest to central Lexington. Lexington was a farming community from colonial times until it reached its peak shortly after 1900, when suburban growth began. The farm produce was sold in Boston's Quincy Market—site of today's Rouse festival shopping development. Suburbanization accelerated after World War II, and 1990 population was near 30,000.

The open triangle of Lexington Green was the site of the first military action of the American Revolution, where minutemen met 700 British regulars. It is about as close to sacred soil as you will find in the United States. Around the green today are several typical colonial residences and a local tavern that was popular at the time of the Revolution. But the stereotypic white-steepled church facing the green is relatively new. The recent expansion of Lexington's business district has diminished the colonial flavor of the place.

From Lexington Green follow Massachusetts Avenue, then Route 2A into the town of Concord. This is the Battle Road, route of the British force after the brief skirmish at Lexington toward its ultimate goal—the store of arms and ammunition at Concord. The road and its immediate borders have been incorporated into the Minute Man National Historical Park, established in the 1960s, with visitor centers and numerous developed sites.

The Battle Road was one of many colonial roads that radiated from Boston to agricultural towns like Lexington and Concord. This route passed through a low gap in the edge of the Boston Basin. Today, heavily traveled Route 2, bypassing both town centers, follows the same corridor.

As you near the Concord town center, follow the signs to Walden Pond. A spot adjoining this classic glacial kettle-hole was home to writer-naturalist-philosopher-American hero Henry David Thoreau (b. 1817; d. 1862) from 1845 to 1847. By accident of ownership the pond has remained almost unchanged since those days, although some recreational use is permitted, especially swimming in summer. Thoreau's stay at Walden Pond has given the place a special mystique, greatly enhanced during the counter-culture movement of the 1960s and 1970s, when Thoreau was the object of tremendous popularity.

Walden Pond serves as a textbook glaciological example for local schools and colleges. Eastern Massachusetts has many similar ponds, but most of them have had substantial and sometimes detrimental development. Two real-estate projects have been recently proposed near Walden Pond. Strong public protest led to abandonment of the schemes and purchase of the site for preservation. Meanwhile, a very large public dump operates nearby.

If you are tempted to stop and stroll around the attractive site, especially on a pleasant summer day, be prepared to be stopped and told to wait a few minutes to a few hours. For environmental reasons, the number of people allowed on the preserve at any one time is strictly limited.

The village of Concord is a little more than a mile north of Walden Pond. Another pocket of level, deep soil (a former glacial lake plain) attracted English settlers to Concord in 1636—the second inland settlement in the Massachusetts Bay Colony. At one time the village was the center of an important asparagus-growing area. The famous Concord grape was perfected here. An early railroad to the American "West" joined Concord to Boston and hauled not only farm produce but also ice, cut commercially from Walden Pond!

Concord attracted a number of important nineteenth-century American writers, poets, artists, and philosophers, including Thoreau. Heavy traffic, busy upscale shops, and the Minuteman Office Center symptomize today's pressure of urbanization, counteracting the strong efforts for preservation, in the once-quaint village. A mile northwest of the village center is the historic North Bridge over the Concord River, where minute men drove back a British attempt to cross the river, firing the "shot heard 'round the world."

Follow Route 62 north from Concord center, along the Concord River valley, to Bedford—another colonial village with many historic buildings—and on, via Route 4, to Billerica. With none of the wealth and glamour of Concord, Billerica has been the scene of extensive modest post-World War II residential development and strip shopping centers.

NINETEENTH-CENTURY INDUSTRIAL SATELLITES

At North Billerica, a small complex of nineteenth-century mill buildings typifies the manufacturing so common across New England at that time. The mills were powered at first by the fall of the Concord River over a small dam, but they turned to coal and more reliable steam power, and finally to electricity before they closed.

Raw materials reached North Billerica initially by barge on the Middlesex Canal, built in 1803, which ran directly through the mill complex. The canal joined the docks at Boston with the upper reaches of the Merrimac River. A parallel railroad completed in 1833 (still visible across the river from our route) soon put the canal out of business. The abandoned bed, overgrown with trees and shrubs, can still be seen at several places along the route.

When the mills closed, the buildings were empty until they came to meet the needs of small, start-up high-technology companies for low-cost office and laboratory space. Many observers credit the large supply of sound, vacant mill structures as an important factor in localizing high-tech industries in New

Restoration of mills and canals in Lowell, Massachusetts, one of the world's greatest nineteenth-century industrial developments.

England. For a variety of historical reasons—among them research at MIT and long-established, entrepreneurial electrical industries—Boston was a center of much early high-tech activity. And much of the industry's critical growth occurred during and immediately after World War II, when construction was severely inhibited.

Take Route 129 from North Billerica to the eastern edge of Chelmsford, then follow the Route 3 freeway north directly into central Lowell, and follow the signs to the parking lot at the Lowell National Historical Park. While the United States has long had a national park system, the concept of an historical park is relatively new. Increasingly, historical parks are being used to restore and preserve urban environments, and Lowell is one case.

In the 1830s, Boston industrial entrepreneurs bought the site of Lowell to dam the Merrimac River and build textile mills. There was only a small settlement there, where a series of locks permitted small craft to pass a series of falls and rapids. Textile manufacturing in New England was experiencing phenomenal growth. Good power sites nearer Boston had become scarce. With railroad transportation, growing capital availability, and improving engineering technology, the time had come to harness the power of this much larger stream. The eventual result was the gigantic mill complex now partly preserved by the national park.

If you didn't get lunch somewhere between Concord and Billerica, the park is a good place to pause. Otherwise, begin your park visit with the short video summarizing the establishment and growth of the city of Lowell. Then take the extensive guided trip through the plants and canals, and study the informative displays in the headquarters building. Don't miss the exhibits in the restored boarding house of the Boott Mills, covering the histories of "mill girls," immigrants, and organized labor.

Lowell mills were an important source of engineering advances and technological inventions in textile manufacturing, which reduced labor requirements, increased productivity, and steadily improved quality. The city itself grew very rapidly throughout the nineteenth century. Factory help came in the early years from nearby farms and villages, mostly young Yankee girls. Later labor needs were met by immigrants from Europe—mainly Ireland and Greece—and French Canada.

Lowell enjoyed nearly a century of economic prosperity until the early 1900s, when the cotton textile industry began to close its northern facilities in favor of the growing capacity in the South. Lowell was extremely hard-hit and was virtually moribund until the 1960s. It was one of the last of New England factory cities to attract post-World War II high-technology industries—in this case, Wang Laboratories, which quickly became Lowell's largest employer.

Meanwhile, many Puerto Rican and Cuban immigrants moved into the city's low-rent districts, soon to be followed by Vietnamese and Cambodians. When the shakeout in the computer industry struck in the late 1980s, the minority community lived in the middle of a much revitalized metropolitan Lowell with more than

40,000 workers in a considerably diversified industrial base. The national park is the core of a small but significant new tourist business, and urban planners and preservationists from many parts of the world visit the place to examine the renovation and adaptation of its old buildings.

THE HOME STRETCH

Leave the Lowell National Historical Park, and head south on Route 3 toward Route 128. As you leave the park area and its uniformed rangers, recall that the Mall in Washington, D.C., is also administered by the National Park Service. In a way, "this is where we came in." In a way, too, virtually everything we've seen on this trip through Megalopolis merits recognition as a part of the American experience, something all of us can profitably observe and understand firsthand, a piece of the country that deserves our attention and reflection.

Return to Boston via Route 128. The unfolding array of electronics, computer, and instrument-company offices tells you why this corridor was once called "America's Technology Highway."

Built in the early 1950s, this was probably the country's first metropolitan circumferential freeway. The timing and location were fortuitous. Central Boston was suffering from severe congestion following the wartime years of great high-tech industrial growth in an old, underbuilt physical plant. The satellite mill towns had a surfeit of empty industrial floor space. Between the inner city and satellite ring, the suburban ring offered abundant attractive residential, office, and laboratory sites.

The Route 128 corridor allowed the merger of all of these unique needs and opportunities and provided an ideal location for the management and engineering functions of high-tech industries. Development was quick and impressive. There followed for some time a widespread, probably erroneous, assumption in the planning and industrial-development fields that the freeway not only localized economic growth but actually caused it. The corridor became a new "outer city," with poor radial linkage to the center, except for the radial rail lines, the Massachusetts Turnpike, and the much more recent Interstate 93.

Exit Route 128 to the Massachusetts Turnpike, and head east through Newton, the Brighton district of Boston, the northern tip of Brookline, and the Back Bay area to central Boston. If you have some spare time, get off the Turnpike and follow the arterials and side streets through these older western suburbs. Beginning in Beacon Hill and Back Bay, this was long the general direction of highest-value residential growth in the metropolitan area. If you don't have a topographic map, oval street patterns and some names on the street map indicate the locations of large drumlins, and you can explore and observe the local effects of elevation on housing values, as well as absorb much more of the character of these neighborhoods. The Turnpike passes beneath the Prudential Center megastructure, on the south edge of Back Bay, before returning you to your downtown destination and the end of the eight-day field trip through Megalopolis.

Coda

I suggest that you now run the entire route in reverse, all the way back to the Washington Monument. You always see things differently going the other direction. And there are many different ways to view Megalopolis!

PART THREE

Resources

△ Hints to the Traveler

The maps in this field guide, though numerous, are small-scale maps that are meant for strategic planning and orientation, not for detailed navigation, especially in the inner cities. You will need a good road atlas and a good street map for each of the metropolitan areas, so that you do not get lost. In addition, it is always a real advantage to bring along the 1:250,000-scale topographic quadrangles of the U.S. Geological Survey, which are the best maps for travel in the United States and whose scale is appropriate for automotive travel. Such maps are readily available on our route in map stores in Washington, D.C., Baltimore, Maryland, and New York City, New York (refer to the "Yellow Pages" of the telephone directory for information). But you can also acquire them in advance from the U.S. Department of the Interior, Geological Survey, Denver Federal Center, Denver, Colorado, 80225, U.S.A.

In addition to having a set of additional maps, arm yourself with local guidebooks and brochures. Although most of these guidebooks tend to be of the "bed and breakfast" type, you can find an adequate selection so that you have a small library with you to help fill in the gaps. Of course the best way to answer questions that inevitably arise on any highway trip is to bring along the 3,000-page *Columbia Encyclopedia* (Columbia University Press). Although it is large, heavy, oversized, and expensive, it is a reference work that every household should own and you will be glad you have it on the trip if your curiosity leads to questions about people, places, and events.

The overnight stops are designed to avoid, as much as possible, the congestion (and, in some cases, the high prices) of the major downtowns on our route. But even in Baltimore, near the city's spectacular Inner Harbor redevelopment, for example, or in Midtown Manhattan, you can do some checking in advance of your arrival to get the very best rates possible. In the tourist off-season, rates are lower than during peak times, and various kinds of discounts are available (senior citizen, weekend, corporate, and the like) if you ask. Reservations are important in some city locations, especially during tourist peaks and in Midtown Manhattan.

It is always hard to predict the weather for a trip that will likely be run by readers at all times of the year, so it is wise to pay attention to the Weather Channel that is available on cable television and to weather predictions in the local newspaper prior to your departure. Summers can be hot and humid, even in New England, and winters can be difficult, even in Washington and Baltimore, so use your best judgment with respect to clothing. Precipitation can occur at any time, so a water-repellent coat is recommended. If you wish to dine at fancy restaurants, then a tie and dinner jacket are required. This is not the Middle West, where comfortable, informal clothing is the norm. And be sure to bring along a pair of binoculars, for there will be many opportunities to see birds along the route, whether we are along the shores of the Chesapeake Bay in Maryland, or Raritan Bay in New Jersey, or Walden Pond in Massachusetts. Also, each of the metropolitan areas we will visit has a rich urban ecology, where birdlife is abundant, and be sure to bring along your running or walking shoes, for there are great opportunities to explore the cities and the countryside on foot.

△ Suggested Readings

Before or during the trip, be sure to acquire the kind of road maps—state, county, metropolitan area, city, and topographic—mentioned in the Hints to the Traveler section. These are absolutely essential. Also, buy local newspapers en route. Real estate, business, and general news sections, especially in Sunday editions, are full of vignettes that provide insight into patterns of development and circulation. Some of these papers are available throughout the region (Megalopolis) and a few, such as the *New York Times,* can be bought just about everywhere in the United States.

During the trip, pick up information brochures on points of interest, cities, counties, and the like, and especially those that contain maps. Also, read the plaques on roadsides, buildings, and monuments. They are unique sources of information about a place which local people think is important.

After the trip, review the items you have checked in the field guide or included in your notes that you want to know more about. Pursue them in the encyclopedia and your library. A lot of people simply do not have the time to keep up with all the reading during the trip, but they want to learn more about a place when they return home. So do take good notes to remind yourself as to what to read when you once again sit in the comfort of your favorite chair and become the venerable armchair tourist.

I thought you also might like to be aware of some general references that might prove useful to you at any time. Obviously there are hundreds of excellent books on the market, and the following list comprises but a sampling. All of the books are relevant to our tour of Megalopolis, and all of them are readable.

For regional physical geography, for this part of the country, the standard work remains Nevin M. Fenneman, *Physiography of Eastern United States* (McGraw-Hill, 1938), Fenneman's *Physiography of Western United States* (McGraw-Hill, 1931) is the companion volume. Also useful is Charles B. Hunt, *Natural Regions of the United States* (W. H. Freeman, 1974), which provides much detail on the new physiography of Megalopolis in its national and natural-science context.

The standard work on which this region is named is Jean Gottmann, *Megalopolis: The Urbanized Northeastern Seaboard of the United States* (The Twentieth Century Fund, 1961). This treatment is broad, insightful, and especially interesting because of its historical and dynamic approach. *Since "Megalopolis": The Urban Writings of Jean Gottmann,* edited by Jean Gottmann and Robert A. Harper (Johns Hopkins University Press, 1990), is a valuable update on Professor Gottmann's thinking about the region and cities, in general.

Four books are of great help for anyone who wants to understand the changing nature and landscapes of urban America since the arrival of the Industrial Revolution in the United States. These are: David Schuyler, *The New Urban Landscape: The Redefinition of City Form in Nineteenth-Century America* (Johns Hopkins University Press, 1986), an acclaimed and highly readable history; Jon C. Teaford, *The Twentieth-Century City: Problem, Promise, and Reality* (Johns Hopkins University Press, 1986), a short history without footnotes; Jon C. Teaford, *The Rough Road to Renaissance: Urban Revitalization in America, 1940–1985* (Johns Hopkins University Press, 1990), a longish account of how cities in the Northeast and Middle West have coped with economic, demographic, and political decline; and Kenneth T. Jackson, *Crabgrass Frontier: The Suburbanization of the United States* (Oxford University Press, 1985), a definitive history. For those who are especially keen to learn more about New York City proper, two books are especially significant: Thomas Bender, *New York Intellect: A History of Intellectual Life in New York City, from 1750 to the Beginnings of Our Own Time* (Alfred A. Knopf, 1987); and Leonard Wallock, editor, *New York: Culture Capital of the World, 1940–1965* (Rizzoli, 1988). A classic study for Boston is Herbert J. Gans, *The Urban Villagers: Group and Class in the Life of Italian-Americans* (The Free Press, 1962). Also notable is Kevin Lynch, *The Image of the City* (The M.I.T. Press, 1960), for his pioneering efforts on how to look at (i.e., interpret) cities. Boston and Jersey City, New Jersey, are two of the three cities featured in the book (Los Angeles is the third). And don't miss reading Angus Gillespie and Michael Rockland, *Looking for America on the New Jersey Turnpike* (Rutgers University

Press, 1989), a first-rate book. Another fine book is Amanda Dargan and Steven Zeitlin, *City Play* (Rutgers University Press, 1990).

For those who are interested in nature writing on the region, you are in luck. From south (Washington, D.C.) to north (Boston), these books include: William W. Warner, *Beautiful Swimmers: Watermen, Crabs, and The Chesapeake Bay* (Little, Brown, 1976), which won a Pulitzer Prize in 1977; Tom Horton, *Bay Country* (Johns Hopkins University Press, 1987), which won high praise in the reviews; Louis J. Halle, *Spring in Washington* (Johns Hopkins University Press, 1988; reprint of the 1947 edition), a renowned classic; John McPhee, *The Pine Barrens* (Farrar, Straus & Giroux, 1968), the book that quite possibly made John McPhee a household name in the United States; Jonathan Berger and John W. Stinton, *Water, Earth, and Fire: Land Use and Environmental Planning in the New Jersey Pine Barrens* (Johns Hopkins University Press, 1985), a little known, but highly acclaimed book on the human and landscape history of this unique region; John Kieran, *Footnotes on Nature* (Doubleday, 1952), a splendid book on the natural life of the New York City region that one only sees on walks; John Kieran, *Natural History of New York City,* second edition (Fordham University Press, 1982), one of the most famous books of nature writing for the East Coast; Hal Borland, *This Hill, This Valley* (Simon and Schuster, 1975), which recounts the cycle of a year on Borland's hill-country farm in northwestern Connecticut, Borland being one of America's most esteemed and most-read nature writers of his time; and a new book, *Nature Walking* (Beacon Press, 1991), by Ralph Waldo Emerson (author of the 1836 essay, "Nature") and Henry David Thoreau (author of the 1862 meditation, "Walking"). Of course, no education is complete without a thorough reading of Thoreau's *Walden Pond,* of which there are numerous editions.

Finally, there are a series of guidebooks that may be of interest to the traveler: *The Smithsonian Guide to Historic America,* a series of twelve volumes organized by regions and published in 1989–1990 by Stewart, Tabori, and Chang; *The States and the Nation,* a series of fifty books published for the national Bicentennial of the American Revolution by W. W. Norton and the American Association for State and Local History; *Rivers of America,* a series of about fifty books published during the 1940s and 1950s by Farrar & Rinehart and Rinhart; and the famous volumes produced by the Works Progress Administration/Federal Writers Project in the later 1930s, some of which have been reissued with new introductions. Check with your local bookstore or library for specific titles within these series that may be of use for you on the trip. *The Smithsonian Guide to Historic America* books are readily available; and Volumes I (*Virginia and the Capital Region*), II (*Southern New England*), and III (*The Mid-Atlantic States*) are key for Megalopolis. Within *Rivers of America,* many of the titles are out of print, but the books of interest are: *The Brandywine* by Henry Seidel Canby; *The Charles* by John Easter Minter; *The Connecticut* by Walter Hard; *The Delaware* by Harry Emerson Wildes; *The Housatonic* by Chard Powers Smith; *The Hudson* by Carl Carmer; *The Potomac* by Frederick Gutheim; *Rivers of the Eastern Shore* by Hulbert Footner; *Salt Rivers of the Massachusetts Shore;* and *Twin Rivers: The Raritan and the Passaic* by Harry Emerson Wildes. For *The States and the Nation* and WPA/FWP books, they are organized by individual states.

Michael Barone and Grant Ujifusa, *Almanac of American Politics* (National Journal, 1990, but published annually), contains concise, insightful narratives on the cultural geography and history for all U.S. Congressional Districts, including those in Megalopolis. They are also accompanied by short statistical profiles on income and demography. Finally, two U.S. Census publications that are basic references for anyone traveling in the United States are: *County and City Data Book* (1988, but published at five-year intervals); and *State and Metropolitan Area Data Book* (1991, but published at five-year intervals). These books contain current and recent data on scores of items concerning population and economy for counties, cities, and metropolitan areas. Believe it or not, they provide interesting reading.

△ Index